T0140273

Environmental Science and Engineering

Series Editors

Ulrich Förstner, Technical University of Hamburg-Harburg, Hamburg, Germany

Wim H. Rulkens, Department of Environmental Technology, Wageningen, The Netherlands

Wim Salomons, Institute for Environmental Studies, University of Amsterdam, Haren, The Netherlands

More information about this series at http://www.springer.com/series/7487

Vishwambhar Prasad Sati

Sustainable Tourism Development in the Himalaya: Constraints and Prospects

 Springer

Vishwambhar Prasad Sati ⓘ
Department of Geography and Resource
Management
School of Earth Sciences
Mizoram University (A Central University)
Aizawl, Mizoram, India

ISSN 1863-5520   ISSN 1863-5539 (electronic)
Environmental Science and Engineering
ISBN 978-3-030-58856-4   ISBN 978-3-030-58854-0 (eBook)
https://doi.org/10.1007/978-3-030-58854-0

This Springer imprint is published by the registered company Springer Nature Switzerland AG
The registered company address is: Gewerbestrasse 11, 6330 Cham, Switzerland

Preface

Tourism, mainly pilgrimage tourism, is a centuries-old practice in the Uttarakhand Himalaya. The description of the highland and river valley pilgrimages is mentioned in the famous religious wisdom of Hinduism such as the Mahabharata and the Ramayana. Historical pieces of evidence depict that the Pandavas while proceeding to the Swarga (heaven) from Uttarakhand constructed the Kedarnath temple during the Dwapar Yuga.[1] Although the Badrinath temple was very ancient, it was renovated by Adi Guru Shankaracharya during the nineteenth century (820 AD). Likewise, the temples of other pilgrimages were constructed during the ancient period, and many of them were renovated by Shankaracharya, such as the temples of Jageshwar and Adi Badri. The Ganga River, a lifeline and a way of life for the people of Northern India, is pious and eminent. It originates and flows from the Uttarakhand Himalaya. The Hindus call it 'The Mother Ganga,' and it is one of the most important pilgrimages. Most of the river valley and the highland pilgrimages are situated along the Ganga River. The word 'Ganga' is pronounced in all the rituals and performed by the people on a day-to-day basis. Its water is pious, life-supporting, and used in all the auspicious occasions from birth to death. It is symbolized as nectar, which provides immortality.

Besides the world-famous pilgrimages, the Uttarakhand Himalaya is bestowed with numerous natural locales for practicing natural and adventure tourism. The landscape is spectacular with varying topographies from the river valleys to the middle altitudes, the highlands, the alpine meadows, and the snow-clad mountain peaks, which are the geographical components of tourism. The national parks and wildlife sanctuaries are the other major eco-tourism destinations. The natural locales include famous summer resorts, where climatic conditions are very feasible during summers. The tourists from the Ganges valley and the rest of India visit these natural locales during the summers to escape themselves from the sunstrokes. Adventure tourism including river-rafting, mountaineering, trekking, and skiing is very popular in the Uttarakhand Himalaya. During the winter season, heavy

---

[1]According to Hindu mythology, there were four Yugas—Satyuga, Tretayuga, Dwaparyuga, and Kalyuga. Now, we are living in Kalyuga.

snowfalls up to the middle altitudes (1600 m); therefore, the tourists visit these destinations. Besides, rural and health tourism are growing areas with enormous prospects.

'Atithi Devo Bhavah,' Guest is God, is the key mantra (belief) of people of the Uttarakhand Himalaya, which they have been following for the centuries. Here, the people are peace-loving, welcome the tourists/pilgrims as God. Further, various cultural components of tourism—pilgrimages, fairs, and festivals—support sustainable tourism development. Folklores—songs and dances, foods and beverages, and art and crafts—are the other prominent aspects of tourism development. Sustainable tourism can be practiced by integrating folk culture with the natural attraction of tourist locales.

The entire Uttarakhand Himalaya is known as the 'Land of Gods and Goddesses' (Dev Bhumi). The mountains, rivers, and forests are named after Gods and Goddess and the fairs and festivals are celebrated to appease them, almost every month. A proverb 'Where folk dances with nature's rhythm' is very popular and befitting to this region. This is also the reason behind the peace-loving nature of the people. The strong background of rich culture and custom further accelerates the high potential of sustainable tourism development of the region.

Although the Uttarakhand Himalaya has plenty of natural and cultural places of tourists/pilgrims' interests, it lags in substantial infrastructural facilities—transportation, accommodation, and institutions—for tourism development. Inaccessibility, remoteness, and fragility of the landscapes are the major hindrances for the development of infrastructural facilities. Roads are traversed only along the river valleys and in the middle altitudes. Further, the road condition is bad. Landslides along the roads are very common, mainly during the rainy season. This leads to severe roadblocks and accidents. Accommodation facilities are not adequate. Tourists/pilgrims face an enormous shortage of accommodation during the peak tourists/pilgrims' seasons. Similarly, institutional facilities for tourism development are yet to be provided. There are several panoramic landscapes situated in the highlands, which have lots of potential for tourism development. However, these areas are unexplored and unpublicized due to lagging institutional facilities.

I spent about 30 years of my early life in the remote part of the Uttarakhand Himalaya and have travelled the entire region several times. While growing up, I noticed beautiful landscape features, rich culture and customs, and poor socioeconomic conditions of the region. Besides, I have also experienced the occurrences of the atmospheric events from time to time, mainly during the monsoon season, which leads to severe catastrophes. The rich natural and cultural components of tourism development in the Uttarakhand Himalaya are not harnessed optimally and thus, it could not receive a progressive position in tourism development. Keeping all these constraints and prospects of tourism development in mind, I decided to write a book on overall aspects of sustainable tourism development in the Uttarakhand Himalaya. Although academic works were carried out on various tourism-related issues in the forms of research papers and articles, no concrete work with high impact on sustainable tourism development in the Uttarakhand Himalaya had been done. This study elaborates on the geographical and cultural components of tourism development scientifically. The description related to tourism development in the Uttarakhand Himalaya is based on the data, collected from the secondary source—Uttarakhand Tourism Development Board, Dehradun. Also, I have been working on different aspects in the environment and development of the Himalaya for the last 30 years. In the tourism aspects, I have already contributed substantially. However, this work is unique because it incorporates all aspects of tourism development in geo-environmental and cultural perspectives. It is a noteworthy work, useful for all the stakeholders who are involved in tourism development such as the policy-makers, academicians, development agents, and the students at different levels. I acknowledge the support of Ms. Vishwani Sati, who edited all the chapters of the book thoroughly. Finally, I dedicate this work to my beloved parents—Late Smt. Saradi Devi Sati and Late Shri. Shiv Dutt Sati for their encouragement, support, and blessings before and after their death.

Chakrata, India   Prof. Vishwambhar Prasad Sati, D.Litt.
January 2020

About This Book

This book, titled *Sustainable Tourism Development in the Himalaya: Constraints and Prospects*, presents insights and a detailed description of tourism development in the Uttarakhand Himalaya. Here, ample geographical and cultural components of tourism support the basis for sustainable tourism development. The snow-clad mountain peaks, the alpine meadows, the highlands, the Middle Himalaya, the Shivalik ranges, the river valleys, and the Doon valley provide unique features and spectacular landscapes. Forest landscapes are panoramic. The rich culture and cultural heritage, supported by fairs and festivals, are among the major tourists/pilgrims' attractions. However, tourism support systems (carrying capacity)—transportation, accommodation, and institutions—are not sufficient. Therefore, tourism has not developed substantially although it shares about 50% of the GSDP.

This book is divided into 12 chapters—Introduction, Geographical and Cultural Components of Tourism, Types of Tourism and Tourist Places, Trends of Tourism, Major Tourism Circuits, Case Studies of Major Tourists/Pilgrims Routes, Infrastructure Facilities for Tourism Development, Homestay Tourism, Tourism Carrying Capacity, Sustainable Tourism Development: Constraints and Prospects, and Conclusions. The chapters are supported by substantial tables, figures, and models. It is unique, first of its kind, a detailed study of all the aspects of tourism development in the Uttarakhand Himalaya. The book is quite beneficial for all stakeholders—students, scholars, academicians, and policy-makers.

Contents

About the Author

Vishwambhar Prasad Sati (b. 1966), Doctor of Letters (2011), and Ph.D. (1992), is a Senior Professor of Geography and Resource Management, Mizoram University, Aizawl, India. He, having a teaching and research experience of above 30 years, has devoted almost all his career years in the development of mountain geography/studies. He has served many national and international educational and scientific institutions in various capacities, such as Associate Professor at 'Eritrea Institute of Technology' Asmara, Eritrea, NE Africa (2005–2007) and Professor in 'Madhya Pradesh Higher Education' (1994–2005 and 2007–2012). He has been a CAS-PIFI Fellow (2016), Visiting Scholar of CAS (2014), Visiting Scholar of TWAS (2010), worked at IMHE, Chengdu, China; Visiting Scholar of INSA (1012), General Fellow of ICSSR (2008–2009), worked at HNBGU, Srinagar Garhwal, an Associate at IIAS, Shimla (2008) and Research Fellow of GBPIHED (1993). He has completed 14 research projects; composed 28 text and reference books; published 110 research papers in journals of international and national repute and many articles in magazines and newspapers, presented research papers in 36 countries and all over India, received fellowships from 38 research organizations to participate in various international events, supervised Ph.D. thesis, organized conferences, chaired many academic sessions, served as a Resource Person in several national and international conferences; and is currently serving many international professional bodies as a member, editor, and reviewer.

Acronyms

ABS	Asan Bird Sanctuary
BJP	Bharatiya Janata Party
BRO	Border Road Organization
BWS	Binsar Wildlife Sanctuary
CECC	Cultural and Economic Carrying Capacity
CNP	Corbett National Park
CWS	Chilla Wildlife Sanctuary
DDUGAHS	Deendayal Upadhyaya Griha Awaas Home Stay
DoES	Directorate of Economics and Statistics
ECC	Environmental Carrying Capacity
EPI	Environmental Performance Index
ESI	Environmental Sustainability Index
GBPIHED	Govind Ballabh Pant Institute of Himalayan Environment and Development
GDP	Gross Domestic Product
GNP	Gangotri National Park
GMVN	Garhwal Mandal Vikas Nigam
GWS	Govind Wildlife Sanctuary
ICC	Institutional Carrying Capacity
INR	Indian Rupees
IT	Information Technology
KMVN	Kumaon Mandal Vikas Nigam
MoT	Ministry of Tourism
NDBR	Nanda Devi Biosphere Reserve
NDRJY	Nanda Devi Raj Jat *Yatra*
NH	National Highway
NHAI	National Highway Authority of India
NHIDCL	National Highway and Infrastructure Development Corporation Limited

PRASHAD	Pilgrimage Rejuvenation and Spiritual, Heritage Augmentation Drive
PWD	Public Work Department
RJNP	Rajaji National Park
SGDP	State Gross Domestic Products
SWOC	Strengths, Weaknesses, Opportunities, and Challenges
TC	Tourism Circuit
TCC	Tourism Carrying Capacity
UNEP	United Nations Environment Programme
UNESCO	United Nations Educational, Scientific and Cultural Organization
UNSDGs	United National Sustainable Development Goals
USD	United State Dollar
USN	Udham Singh Nagar
UTDB	Uttarakhand Tourism Development Board
VFNP	Valley of Flowers National Park
WTO	World Tourism Organization

List of Figures

List of Tables

Chapter 1
Introduction

Tourism is a process in which a substantial amount of time is spent on leisure, pilgrimage, adventure, and for commerce, away from home. A social, cultural, and economic phenomenon, tourism entails the movement within or outside the country for pleasure, pilgrimage, and business. WTO (2008) defines tourism as activities of people traveling within and outside of their usual environment, not for more than a year for leisure, business, and other purposes. Travel and tourism have become prominent smokeless industries and growing economic sectors all over the world, resulting in the enhancement of income and economy, and augmented employment. Along with an increase in infrastructural facilities, tourism activities have increased manifold.

Tourism is the world's third-largest export category, following chemicals and fuels, and followed by automotive products and food (WTO 2017). It is called a smokeless industry. Tourism has been included in the United Nations Sustainable Development Goals (SDGs) No. 8, 12, and 14. The World Tourism Organization (WTO) aims to promote tourism for economic growth, inclusive development, and environmental sustainability (UNWTO 2014). Travel and tourism accounted for USD 8.9 trillion in contributions to the world's GDP, which is 10.3% of the global GDP. It has provided 330 million jobs, 1 in 10 jobs around the world. World tourist arrival is expected to be 1.56 billion by the year 2020. In 2018, international tourist arrivals grew 5%, currently being 1.4 billion. It demonstrates a growth rate of 4.1% from 1995 to 2020. The 22 destination countries of the Indian Ocean, including India, are forecasted to receive 179 million international tourists with a 6.3% growth rate for 1995–2020 (WTO 2020).

In 2018, income generated from tourism grew to be about 1.7 trillion USD. Tourism also manifested in the creation of jobs. World Gross Development Products (GDPs) increased by 3.6% during the same time. In comparison to 2017, about 121 billion USD extra revenues from international tourism were generated in 2018 (WTO 2019). Globally, Luxembourg, followed by Australia, has earned about USD

© The Editor(s) (if applicable) and The Author(s), under exclusive license to Springer Nature Switzerland AG 2020
V. P. Sati, *Sustainable Tourism Development in the Himalaya: Constraints and Prospects*, Environmental Science and Engineering,
https://doi.org/10.1007/978-3-030-58854-0_1

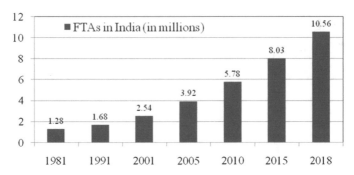

Fig. 1.1 Foreign tourist arrivals in India (1981–2018). *Source* Bureau of immigration, Government of India

4,900 per international arrival of tourists. India is one of the important destinations, obtained a 19% increase in tourism arrival and a 10% increase in tourism receipts. It jumped from 40th rank to 34th rank in the world's total number of tourists' arrival in 2019 and got included in the list of the top 35 lower-middle-income countries in improving its overall score in tourists' arrival.

Tourism contributed about 247 billion USD to the country's (India) GDP in 2018, which is 9.6% of the total GDP. Ministry of Tourism (MoT) (2019) reported that about 10.89 million tourists visited India in 2019, resulting in a growth of 3.2%. The number of 1854 million domestic tourists that visited India in 2018 was about 1854 million, which was 1657 million in 2017, resulting in the growth rate being 11.9%. The income generated from domestic tourism in 2019 was Rs. 2,109,810 million, with a growth of 8.3%. The figure was 29.962 million USD, with a growth of 4.8% from foreign tourists. The two major schemes launched by the Ministry of Tourism for creating tourism infrastructure in 2014–2015 were Swadesh Darshan—Integrated Development of Theme-based Tourism Circuits and PRASHAD—Pilgrimage Rejuvenation and Spiritual, Heritage Augmentation Drive. Under the scheme, 15 thematic circuits were identified for tourism development, viz. North-East India Circuit, Buddhist Circuit, Himalayan Circuit, Coastal Circuit, Krishna Circuit, Desert Circuit, Tribal Circuit, Eco Circuit, Wildlife Circuit, Rural Circuit, Spiritual Circuit, Ramayana Circuit, Heritage Circuit, Tirthankar Circuit, and Sufi Circuit. 51 sites of 28 states were identified under the PRASHAD scheme for tourism development (MoT 2019).

India shared 1.2% international tourist arrivals in 2018 and was ranked 7th. Foreign tourist arrivals in India from 1981 to 2018 have constantly increased, 1.28 million in 1981 to 10.56 million in 2018. The figure now has increased 10 times, which shows that India has enormous potential for tourism development (Fig. 1.1). The growth percentage of foreign tourist arrivals increased from 31.25% in 1981–1991 to 803% in 2015–2018.

1.1 Sustainable Tourism Development

The Uttarakhand Himalaya has two distinct geo-environmental and cultural enti-ties—the Garhwal Himalaya and the Kumaon Himalaya. The landscape of the Garhwal Himalaya is more fragile, steep, and rugged than the Kumaon Himalaya, which has gentle slopes and fertile valleys. Further, culture varies in both entities. The Garhwal Himalaya is known for its highlands and river valley pilgrimages whereas the Kumaon Himalaya has beautiful summer resorts. There are seven districts in the Garhwal division—Chamoli, Pauri, Rudraprayag, Tehri, Uttarkashi, Dehradun, and Haridwar. The Kumaon division has six districts—Pithoragarh, Champawat, Bageshwar, Almora, Nainital, and Udham Singh Nagar (USN). The total area of the Uttarakhand Himalaya is 53,483 km^2. It has two international boundaries—Tibet (China) in the north and Nepal in the east and it is bordered by Himachal Pradesh from the west and Uttar Pradesh from the south (Fig. 1.2).

Himalaya is believed to be an embodiment of Lord Shiva. It is a center of spiritual attainment for the Hindus, Jains, and Buddhists. The Mount Kailash itself is a center of belief and the Kailash-Mansarovar *Yatra* (procession) is one of the spiritual proces-sions performed by the followers of Hinduism, Jainism, and Buddhism. Pilgrimage to Himalaya is a centuries-old practice (Grotzbach 1994). The highlands and the river valleys pilgrimages in Himalaya (India, Nepal, and Bhutan) are world-famous. Similarly, in the Uttarakhand Himalaya, many spiritual *Yatras* are performed by the

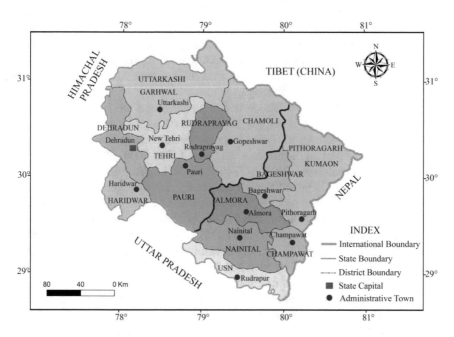

Fig. 1.2 Location map of Uttarakhand Himalaya showing Garhwal and Kumaon divisions

folks. The Nanda Devi Raj Jat *Yatra* (NDRRJ) is very famous and one of the longest *Yatras.*

The Uttarakhand Himalaya, an integral part of Himalaya, has snow-clad mountain peaks; panoramic landscapes—the river valleys, the middle altitudes, the highlands, and alpine pasturelands—rich faunal and floral diversity, rich traditional culture, and cultural heritage, customs, and sacred places—highlands and river valleys. It's visited by exodus pilgrims and tourists drawn from the Indian subcontinent and abroad every year. All types of tourism—natural, cultural, spiritual, park and wildlife, eco-tourism, and adventure are practiced here. It was acknowledged *as* Switzerland of India by Mahatma Gandhi after his first visit to Uttarakhand. The pilgrims believe Uttarakhand as the 'Land of the Gods and Goddesses'. The Ganga, which originates and flows from the Uttarakhand, is called the mother Ganga, which is one of the most pious rivers and a major attraction for the pilgrims. The development of tourism in the Uttarakhand Himalaya also depends on the native people who are very humble and believe in the Guest being God (Atithi Devo Bhavah). Also, the native people are peace-loving and welcoming in nature.

Uttarakhand has the world-famous national parks and wildlife sanctuaries—the Corbett National Park (CNP), the Raja Ji National Park (RJNP), the Nanda Devi Biosphere Reserve (NDBR), the Binsar Wildlife Sanctuary (BWS), the Govind Wildlife Sanctuary (GWS), the Asan Bird Sanctuary (ABS), the Valley of Flowers National Park (VFNP), the Gangotri National Park (GNP); natural locales—Dehradun, Mussoorie, Chakrata, Gwaldam, Nainital, Ranikhet, Almora, and Kausani; the highlands and the river valleys pilgrimages—Badrinath, Kedarnath, Yamunotri, Gangotri, Rishikesh, Haridwar, and many other places of tourist and pilgrim's interests. Because of the diverse nature of tourist places, natural tourism, cultural tourism (pilgrimages), park and wildlife tourism, eco-tourism, and adventure tourism—trekking, mountaineering, river-rafting, water sports, and skiing are practiced in the Uttarakhand Himalaya.

Although the Uttarakhand Himalaya has plenty of the world-famous tourists' places and pilgrimages yet, it has not got an impressive position in tourism-related development. The entire region is lagging in infrastructural facilities such as transportation, accommodation, institutions, and health (Sati 2004, 2018). The Uttarakhand Tourism Development Board (UTDB) is a state government agency, which came into force in 2001 after Uttarakhand got statehood by getting carved out of Uttar Pradesh. Its role is to provide basic amenities to tourists/pilgrims however, it has failed to do any remarkable work in this regard. The two development agencies of the state government—the Garhwal Mandal Vikas Nigam (GMVN) and the Kumaon Mandal Vikas Nigam (KMVN) are also providing services for tourism development. GMVN is headquartered in Dehradun and KMVN is headquartered in Nainital.

The Uttarakhand Himalaya has tremendous potential for tourism development as it has numerous world-famous tourism locales (Sati 2013, 2015, 2018). Tourism is a large industry in Uttarakhand as it generates 50% revenues of the State Gross Domestic Products (SDGPs). Out of the total tourists/pilgrims who arrive in Uttarakhand, 80% of them visit the four pilgrimages. During the last 18 years (2000–2018), total tourists/pilgrims who visited Uttarakhand were 674.35 million of which,

300 million domestic pilgrims and 0.48 million foreign pilgrims visited the major pilgrimages. In Haridwar itself, about 240 million domestic and 0.32 million foreign pilgrims visited. About 56.1 million domestic tourists and 0.5 million foreigners visited the natural locales and 76 million domestic and 0.95 foreign tourists visited the administrative tourist places.

The socio-political boundary of the Uttarakhand Himalaya is full of Hindu shrines belonging to various sects and traditions, which have been centers of pilgrimage for time immemorial. Following the Vedic period, the practice of religious tourism or pilgrimage seems to have gained popularity as evident from the great epic Mahabharata (Bhardwaj 1973). The Uttarakhand Himalaya has immense potential for tourism development as it is bestowed with religious, natural, adventure, and wildlife tourism destinations. Besides the major tourist places of Uttarakhand, the government has introduced 21 new potential tourist destinations for further development. These destinations are Jageshwar, Baijnath, Saat Taal, Bhimtal, Patal Bhuvaneshwar, Chakori, Someshwar, Pithoragarh, Chakrata, Roopkund, Hanol, Asan Barrage, Harshil, Dhanaulti, and Dayara Bugyal. Jauljivi in Kumaon and the Tons River in Garhwal are planned to be developed as adventure tourism destinations. "Udan' scheme was launched in January 2018 to connect the major tourist places of Uttarakhand by air (IBEF 2018).

The first Tourism Policy was introduced in 2001 under the aegis of UTDB. The main objective of the tourism policy was to develop tourist destinations and make them world-class by providing infrastructural facilities. Its purpose was to harness the huge tourism potential of the state by engaging private sector agencies. Income generation, employment augmentation, sustainable growth of tourism products, and integrating tourism for community development were other objectives. NITI Aayog (2018) has proposed methodological framework for sustainable tourism development in the Uttarakhand Himalaya by identifying gaps and best initiatives related to sustainable tourism, state-level dialogues among the stakeholders, integration of sustainable tourism policies, identifying policy, financial, and institutional initiatives to support tourism, and finally development of a regulatory framework for contributing in the development of sustainable tourism.

This book examines the nature and trends of tourists/pilgrims' inflow in the Uttarakhand Himalaya. It also focuses on geographical and cultural components of tourism, tourism circuits, and case studies of some important tourism routes, development of infrastructural facilities for tourism development, tourism carrying capacity (TCC), and the major challenges and opportunities of tourism development in the Uttarakhand Himalaya. Finally, it outlines to harness the natural and cultural carrying capacity of tourism by providing and developing infrastructural facilities.

1.2 Potential of Theme-Based Tourism

The Uttarakhand Himalaya practices several types of tourism such as adventure—trekking, mountaineering, river rafting, and skiing; pilgrimage tourism—highlands and the valleys; cultural tourism; natural tourism; park and wildlife tourism (eco-tourism); and health and rural tourism (Table 1.1). Many natural and cultural locales have been declared as heritage regions, which provide a suitable base for sustainable tourism in Uttarakhand. Along with natural and cultural heritage, there are some historical monuments, which are important sites for tourists. Park and wildlife tourism have enormous potential for practicing ecotourism. Diversity in faunal and floral species is high. They are the major attractions for any nature lover. Heritage tourism, which includes natural heritage such as fauna, flora, landscapes—rivers, lakes, and mountains and cultural heritage such as historical monuments, paintings, sculptures, temples, rites, fairs, and festivals, *attracts tourists in the majority*. Trekking, mountaineering, river rafting, and skiing come under adventure tourism practiced in Uttarakhand. Similarly, cultural tourism is centuries old. Health and rural tourism also have enormous potential. The Uttarakhand Himalaya has a range of climatic conditions. The climate is very feasible during the summer season, mainly in the Middle Himalaya. The hill resorts in the Middle Himalaya can be the destination for health tourism. Besides, the Environmental Sustainability Index (ESI) of Uttarakhand is more than 80%. Further, Uttarakhand tops in Environmental Performance Index (EPI), as it scored 0.8123, which is the highest among the list of best-performing States and Union Territories. The tourists and pilgrims can improve their health after visiting Uttarakhand for a certain period.

1.3 Principles of Sustainable Tourism

Tourism has become a smokeless industry. Its dimension is increasing year by year worldwide. Most of the tourism is practiced in natural areas, which are full of faunal and floral resources. However, increasing tourism is a threat to biodiversity. Keeping environmental degradation and biodiversity loss in mind, sustainable tourism development is completely indispensable. The major challenge of sustainable tourism is to make the optimum use of natural resources for positive tourism.

Sustainable form of tourism can be attained when all the natural, cultural, and human resources are adequately considered and used optimally from a holistic perspective (Dodds 2007). Sustainable tourism is a kind of tourism that considers the current and future impact of tourism on the economy, society, and the environment and addresses these issues sustainably (UNEP/WTO 2005). Sustainable tourism deals with the optimum use of environmental resources and conserves the socio-cultural characterization of the local communities. A sustainable tourism manifests sound economy, the well-being of the locales, nature's conservation and protection of the resources, healthy culture, and satisfaction of guests (Muller 1994; Hunter 1997). Besides, sustainable tourism is referred to as the management and development of tourism. It preserves natural, economic, and socio-cultural integrity, and conserves natural and socio-cultural resources (Niedziolka 2012). All these definitions indicate that sustainable tourism is associated with economic development, and socio-cultural

Table 1.1 Potential of theme-based tourism in Uttarakhand

Themes		Tourism destinations
Adventure tourism	Trekking	Kedartal Trek, Rupin Pass Trek, Kedarkantha Trek, Bali Pass Trek, Auden's Col Trek, Har Ki Dun Trek, Satopanth Lake Trek, The Valley of Flower Trek, Brahmatal Trek, Kalindi Khal Trek, Roopkund Trek, Pindari Glacier Trek, Kafni Glacier Trek, Panchchuli Base Camp, Sunderdhunga Trek, Nanda Devi East Base Camp, Namik Glacier Trek, Gaumukh, Sinla Pass Trekking, Chhota Kailash Trek, Kailash-Mansarovar trek, Bedni Bugyal, Ali Bugyal, and Dayara Bugyal Treks
	Mountaineering	Nanda Devi, Gaumukh, Panchchuli, Om Parvat, Trishul, Chaukhamba, and Bhagirathi Peak
	river rafting	Rishikesh, Shivpuri, Kodiyala, Byasi, Devprayag (Ganga river), Mori (Tons river), Kali, Saryu, and Ramganga Rivers; Boating in Naini Lake, Bhimtal, and Naukuchiyatal
	Skiing	Auli and Chopta
Pilgrimage tourism		Yamunotri, Gangotri, Panch Kedar—Kedarnath, Tungnath, Madhyamaheshwar, Rudranath, and Kalpeshwar; Panch Badri—Badrinath (Badri Vishal), Bhavishya Badri, Adi Badri, Yogadhyan Badri, and Bridha Badri; Rishikesh and Haridwar, Panch Prayag—Vishnuprayag, Nandprayag, Karnprayag, Rudraprayag, and Devprayag; Hemkund Sahib, Pandukeshwar, Jageshwar, Dunagiri, Purnagiri, Jauljivi, and Hanol, Chhota Kailash, and Kailash-Mansarovar
Cultural tourism		Nanda Devi Rajjat, Pandav Nritya, Baishakhi, All Sakrantis, Navratras, Uttarayani, Magh Mela, and Kumbh Mela
Natural tourism		Dehradun, Mussoorie, Chakrata, Gopeshwar, Pauri, Nainital, Almora, Champawat, Bageshwar, Ranikhet, Kausani, New Tehri, Joshimath, Pithoragarh, and Dharchula
National parks and wildlife sanctuaries tourism (Eco-tourism)		RJNP, CNP, GWS, BWS, NDBR, VFNP, and GNP

(continued)

Table 1.1 (continued)

Themes	Tourism destinations
Health and rural tourism	Munsiyari, Chopta, Gwaldom, Jakholi, Kapkot, Lansdowne, Tiuni, Purola, Dwarahat, and Bhararisain

Source By author

and environmental conservation (Creaco and Querini 2003; Muhanna 2006; Richins 2009; Patterson 2016).

UNWTO (2004) defines sustainable tourism as the development that meets the needs of present tourists and host regions while protecting and enhancing opportunities for the future. It is envisaged as leading to management of all resources in such a way that economic, social, and aesthetic needs can be fulfilled while maintaining cultural integrity, essential ecological processes, biological diversity, and life support system. TCC and destination development is a key aspect of sustainable tourism practices, indicating the maximum number of tourist visiting a tourist destination without destroying the physical, economic, and socio-cultural environment (PAP/RAC 1997). Sustainable tourism development makes optimal use of biodiversity resources, maintaining ecology, and conserving natural and cultural environments. It generates income, augments employment, promotes local culture and customs, and contributes to poverty alleviation (WTO and UNEP 2005). UNEP (2012) has developed a new concept of 'Tourism in the Green Economy', which deals with sustainable tourism development.

Tourism demands enormous use of natural resources and it has a tremendous impact on the environment, ecosystems, economy, society, and culture. Tourism contributes effectively to rural and regional development if managed sustainably. Otherwise, tourism leads to the devastation of nature and culture. Tourism broadly depends on national, regional, and local resources and traditions, culture, and customs. Further, tourism largely depends on human resources at all levels mainly on the quality of service providers. Local food and beverages, arts and crafts, and fairs and festivals are the other important products of tourism.

Sustainable tourism development is environmentally sound, economically viable, and culturally acceptable and equitable. It takes care of the fragile environment of the tourism destinations, mainly in the mountainous regions, which are ecologically fragile. Sustainable tourism development is a long term policy planning. It involves economic and infrastructural development, social well-being, and cultural strengthening. Further, it assists in livelihood enhancement through income generation and employment augmentation. It provides a base for optimum use of natural and cultural resources for economic development. It also ensures natural resources conservation and cultural values preservation. Through sustainable tourism, quality tourism products are offered, which are drivers of economic development. Further, it enhances quality transportation, accommodation, and local food and beverages.

A sustainable tourism development model has been developed (Fig. 1.3). The model has three components—economic viability, environmental suitability, and

Fig. 1.3 A sustainable tourism model; *Source* By author

social acceptability. Further, each component has several sub-components, which decide the sustainable tourism development in the Uttarakhand Himalaya. The Uttarakhand Himalaya is economically underdeveloped. The infrastructural facilities for tourism development—transportation, accommodation, and institutions along with health facilities, are lagging. For sustainable tourism development, these facilities can be developed and provided to the tourists and pilgrims. The natural and cultural environment of the Uttarakhand Himalaya is quite suitable for tourism development. However, the ecology is fragile and the landscape is vulnerable. Tourism development also leads to environmental degradation in the forms of depletion of forests, erosion, landslides, and mass movements. Pollution and dumping of waste are other environmental problems. Conservation of forests and landscapes and pollution-free environment and waste management may lead to sustainable tourism development. Social acceptability of tourism is important because tourism leads to cultural erosion. Conservation of culture, customs, and rituals, and performance of fairs and festivals are equally important.

1.4 Impact of Tourism

Tourism is an important sector of economic development and it has a relationship with economic growth and other economic activities (Zhang 2015). Tourism has manifested improvements in transport, changing lifestyles, increased leisure time, international openness and globalization, increased level of education, information and

communication technology, destination development, and improved tourism infrastructure (Matias et al. 2007; Aziri and Nedelea 2013). Tourism has also contributed
to improving rural livelihoods and the quality of life of rural communities (Andereck
and Nyaupane 2011; Sharpley 2002). It has a multiplier impact on income, economy,
environment, and culture. On one hand, it has a positive impact on income and
economy, on the other, it harms the environment and culture. Its impact depends
on the nature and number of tourists/pilgrims and the nature of tourist places and
pilgrimages. The negative impacts need to be managed through a sustainable and
holistic development approach.

1.5   Economic Impact of Tourism

Tourism generates income, augments employment, and enhances livelihoods through
the construction of accommodation facilities and the establishment of economic
avenues. However, developed countries are more benefited by this than the developing
ones. Tourism development is associated with the development of infrastructural
facilities. Better the infrastructural facilities, more substantial is tourism development
and vice-versa.

1.6   Socio-cultural Impact of Tourism

Tourism has both positive and negative socio-cultural impacts. While on one hand,
tourism promotes culture and strengthens it, on the other, it degrades culture. The
local people always try to adapt to the culture of tourists/pilgrims and thus their own
culture degrades. Tourism also frequently leads to changes in the values and behavior
of the local community.

1.7   Environmental Impact of Tourism

Environmental impacts of tourism are enormous in the forms of land degradation,
forest depletion, and air and water pollution. It not only impacts the natural environment, but also impacts cultivated land and ecology at local, regional, and global
levels. Increasing global traffic has led to climate change and global warming and
the role of tourism is noteworthy.

1.8 Scope of the Study

Sustainable tourism development is a very comprehensive and important aspect of the development of the Himalayan region. This study deals with the sustainable tourism development with special reference to the Uttarakhand Himalaya. A detailed description of introduction to sustainable tourism development, the components of tourism—geographical and cultural, types of tourism and major tourists/pilgrims' centers, the trends of tourism and tourists/pilgrims' inflow, major tourists/pilgrims's circuits, case studies of major tourism routes, development of infrastructural facilities for sustainable tourism, sustainable tourism development: constraints and prospects, and conclusions in respects of the Uttarakhand Himalaya are presented in this study.

1.9 The Need and Purpose of the Study

Although the Uttarakhand Himalaya has tremendous potential for tourism development, still tourism development here could not take as good shape as it has in some countries with attractive tourist destinations. Though substantial works have been conducted in the academic field by scholars on the tourism development of the Uttarakhand Himalaya, yet the works do not cover all aspects of tourism development, particularly the geographical perspectives. The major purpose of this study was to examine the potential of sustainable tourism development in the Uttarakhand Himalaya. It described the major constraints and prospects of tourism development. The geographical and cultural components of tourism, nature, and types of tourism, tourism trends, homestay tourism, infrastructural facilities, major tourists' circuits, and case study of major tourism routes were described. This book is unique because it presents a holistic and comprehensive description of tourism development in the Uttarakhand Himalaya for the first time. All three methods—qualitative, quantitative, and descriptive were employed, which further makes the contents and text rich.

1.10 Methodology

This study employed all qualitative, quantitative, and descriptive approaches. An observation method was used to elaborate on tourism activities and the role of tourism in the sustainable development of the Uttarakhand Himalaya. Data were gathered from secondary sources, and through wide and rapid visits to tourist places and pilgrimages. UTDB, Dehradun was the main source of data on tourists/pilgrims' inflow in the tourist places and pilgrimages of Uttarakhand from 2000 to 2018. Data on tourism infrastructural facilities—transportation, accommodation, and institutional, including health facilities, were gathered from the Directorate of Economics and Statistics (DoES), Government of Uttarakhand, Dehradun. Types of tourism and tourist places, trends of tourism and tourists/pilgrims' inflow, tourism circuits, and development of infrastructural facilities were described and analyzed. Data were shown using tables, graphic presentations, and mapping. Sustainable tourism model was developed. The author has visited the tourist places and pilgrimage of the Uttarakhand Himalaya several times. Case studies of three tourist/pilgrim routes, which cover the entire study region, were carried out. The author has observed and described the availability and quality of infrastructural facilities, geographical and cultural components of tourism, and the impact of tourism on environment, economy, and culture. The major constraints and prospects of tourism development have been described, and SWOC analysis was carried out by the author based on his experiences and available data on tourist places/pilgrimages and available tourism infrastructural facilities. The author has divided the Uttarakhand Himalaya into tourism circuits first time. These circuits are based on the nature and location of tourist places/pilgrimages. Concrete and comprehensive suggestions are given for sustainable tourism development in the Uttarakhand Himalaya.

1.11   Organization of the Study

This book comprises 12 chapters along with a preface, list of tables and figures, and acronyms and abbreviations. The first chapter is an introduction. A comprehensive description on the current trends of tourism *in India and the world*; sustainable tourism development in the Uttarakhand Himalaya; the potential of theme-based tourism; principles of sustainable tourism along with sustainable tourism model; the impact of tourism on the economy; socio-culture, and environment, the scope of the study; the need and objectives of the study; and methodology is presented. The second chapter is devoted to geographical components of tourism development. It comprises of an introduction, geographical components—spectacular landscapes, river topographies, forest landscapes, feasible climate—major mountain peaks, major river systems, major glaciers, and conclusions. Cultural components of tourism development are described in the third chapter. This chapter is composed of an introduction, pilgrimages—highlands and river valleys—fairs and festivals, traditional food and beverages, arts and crafts, woolen cloths, and conclusions. The fourth chapter studies the types of tourism and major tourists/pilgrims centers. The important subheadings of the chapter are an introduction, natural tourism—major natural locales and administrative towns/cities—wildlife/park tourism, pilgrimage, and cultural tourism, fairs and festivals, adventure tourism—trekking, mountaineering, river rafting, skiing—and finally conclusions. The fifth chapter deals with the trends of tourism and tourists/pilgrims' inflow. The chapter describes introduction, trends of tourism and tourists/pilgrims' inflow, domestic tourists/pilgrims' inflow, foreign tourists/pilgrims' inflow, domestic pilgrims' inflow in the highland pilgrimages, foreign pilgrims' inflow in the highland pilgrimages, domestic tourists' inflow in the major natural locales/administrative towns, foreign tourists' inflow in the major natural locales/administrative towns, domestic pilgrims' inflow in the river valleys pilgrimages, foreign pilgrims' inflow in the river valleys pilgrimages, analysis of tourists/pilgrims' inflow using descriptive statistics, discussion and conclusions. Chapter six presents major tourists/pilgrims' circuits. The chapter includes an introduction, major tourists/pilgrims' circuits—Uttarkashi, Yamunotri, Harsil, and Gangotri circuit; Dehradun, Mussoorie, Chakrata, Rishikesh, and Haridwar circuit; Panch Badri circuit; Panch Kedar circuit; Devprayag, Kunjapuri, Chandrabadni, Surkanda, and New Tehri circuit; Chilla, Kotdwar, Lansdowne, and Pauri circuit; Ranikhet, Almora, Jageshwar, Kausani, Baijnath, and Bageshwar circuit; Mukteshwar, Nainital, Bhimtal, CNP, and Rudrapur circuit; Champawat, Pithoragarh, Dharchula, and Munsiyari circuit; domestic tourists/pilgrims' inflow in tourism circuit; foreign tourists/pilgrims' inflow in tourism circuit; and conclusions. Case studies of major tourism routes are described in the seventh chapter. In this chapter introduction, Karnprayag, Gwaldom, Almora, Nainital, Ranikhet, and Gairsain route; Kalsi, Chakrata, Tiuni, Hanol, Mori, and Purola route; Rudraprayag, Kedarnath, Madhyamaheshwar, Tungnath, Rudranath, and Kalpeshwar route; and conclusions are described. The development of infrastructural facilities for tourism-related activities is described in Chap. 8. It includes an introduction, transportation facilities—airways, railways, roadways, major national highways, and all-weather roads; accommodation—accommodation units in the major natural and cultural places and classification of accommodation units; institutional facilities including medical and educational; and conclusions. Chapter 9 illustrates sustainable homestay tourism and its prospects. It comprises of introduction, homestay tourism policy in Uttarakhand, district-wise registered units and income from homestay tourism in urban areas, district-wise registered units and income from homestay tourism in rural areas, statistical analysis of homestay tourism in urban and rural areas, increasing homestay facilities, income obtained from homestay tourism, sustainable homestay tourism, the impact of homestay tourism, discussion, and conclusion. Chapter 10 explains tourism carrying capacity and destination development in which a case study of five tourist places has been conducted. The destination development of tourist places has been elaborated. The 11th chapter bears sustainable tourism development: constraints and prospects. The major constraints—natural hazards and disasters, unplanned infrastructural development, unplanned tourism development, mismanagement during the peak pilgrimage seasons, lack of parking facilities, bad quality roads, potable water problem, sanitation problem, solid waste management problem, lack of clean and adequate public convenience facilities, lack of accommodation, lack of health facilities, and lack of trained guides are systematically illustrated. Further, prospects of sustainable tourism development in the Uttarakhand Himalaya such as world-famous pilgrimages, rich culture and customs, world-class summer resorts, spectacular landscapes, varied and suitable climatic conditions, and welcoming behavior of local people are illustrated in this chapter. The last chapter, Chap. 12, is about the conclusions, in which suggestions are given for sustainable tourism development in the Uttarakhand Himalaya.

References

Andereck KL, Nyaupane GP (2011) Exploring the nature of tourism and quality of life perceptions among residents. J Travel Res 50(3):248–260

Aziri B, Nedelea A (2013) Business strategies in tourism. EcoForum J 1(2):5–11

Bhardwaj SM (1973) Hindu places of Pilgrimage in India: a study in cultural geography. Thomson Press, Delhi

Creaco S, Querini G (2003) The role of tourism in sustainable economic development. In: Presented at the 43rd congress of the European regional science association. Jyvaskyla, Finland, 27–30

Dodds R (2007) Sustainable tourism and policy implementation: lessons from the case of Calvia. Spain Curr Issues Tour 10:296–322

Grotzbach E (1994) Hindu-Heiligtumer als Pilgerziele in Hochhimalaya, Erdkunde 48(3)

Hunter C (1997) Sustainable tourism as an adaptive paradigm. Ann Tour Res 24:850–867

IBEF (2018) Uttarakhand: the spiritual sovereign of India. India Brand Equity Foundation, New Delhi India

Karar A (2010) Impact of pilgrim tourism at Haridwar. Anthropologist 12(2):99–105

Matias A, Nijkamp P, Neto P (2007) Advances in modern tourism research: economic perspectives. PhysicaVerlag, Heidleberg

MoT (2019) Ministry of Tourism and Government of India, New Delhi

Muhanna E (2006) Sustainable tourism development and environmental management for developing countries. Probl Perspect Manag 4:14–30

Muller H (1994) The thorny path to sustainable tourism development. J Sustain Tour 2:131–136

Niedziolka I (2012) Sustainable tourism development. Reg Form Dev Stud 8:157–166

NITI Aayog (2018) Report of working group II sustainable tourism in the Indian Himalayan Region, New Delhi

PAP/RAC (1997) Guidelines for mapping and measurement of rainfall-induced erosion processes in the mediterranean coastal areas. PAP-8/PP/GL.1. Split, Priority Actions Programme Regional Activity Centre (PAM/PNUE), in Collaboration with FAO, xii+72

Patterson C (2016) Sustainable tourism: business development, operations, and management; human kinetics: champaign. IL, USA

Richins H (2009) Environmental, cultural, economic and socio-community sustainability: a framework for sustainable tourism in resort destinations. Environ Dev Sustain 11:785–800

Sati VP (2019) Himalaya on the threshold of change. Springer International Publishers, Switzerland, p 250. https://doi.org/10.1007/978-3-030-14180-6

Sati VP (2018) Carrying capacity analysis and destination development: a case study of Gangotri tourists/pilgrims' circuit in the Himalaya. Asia Pacific J Tour Res 23(3):312–322. https://doi.org/10.1080/10941665.2018.1433220

Sati VP (2015) Pilgrimage tourism in mountain regions: socio-economic implications in the Garhwal Himalaya. South Asian J Tour Herit 8(1):164–182

Sati VP (2013) Tourism practices and approaches for its development in the Uttarakhand Himalaya India. J Tour Challenges Trends 6(1):97–112

Sati VP (2004) Uttaranchal: dilemma of plenties and scarcities, New Delhi, published by Mittal Publications

Sharpley R (2002) Rural tourism and the challenge of tourism diversification: the case of Cyprus. Tour Manage 23(3):233–244

UNWTO (2014) Tourism highlights, 2014 Edition, www.unwto.org/pub

UNEP (2012) T-21 model, towards a green economy, UNEP

UNWTO (2004) Indicators of sustainable development for tourism destinations. A guidebook. Madrid

UNEP/WTO (2005a, b) Making tourism more sustainable: a guide for policy makers; united nations environment programme/world tourism organization. Paris, France, Madrid, Spain

WTO (2020) Tourism vision 2020, Capitan Haya 42, 28020 Madrid, Spain

WTO (2019) International tourism highlights 2019 Edition, Capitan Haya 42, 28020 Madrid, Spain

WTO (2017) International tourism highlights 2017 Edition, Capitan Haya 42, 28020 Madrid, Spain

WTO. (2008). Understanding tourism: basic glossary. United Nations World Tourism Organization. https://media.unwto.org/en/content/understanding-tourism-basic-glossary

Zhang WB (2015) Tourism, trade and wealth accumulation with endogenous income and wealth distribution among countries. EcoForum J 1(6):7–13

Chapter 2
Geographical Components of Tourism Development

2.1 Introduction

Tourism is an important human activity, which deserves the praise and encouragement of all people and all governments. It consists of activities—economic, cultural, and leisure and it is related to entry, stay, and movement of tourists from one place to another. In other words, tourism is the sum of the total phenomenon and relationship, related to travel and the stay of tourists without earning money. Geographical components provide suitable bases for tourism development. These components are accessibility and location, space, scenery (landforms, water, and vegetation), and climate (Robinson 1976). The locale is an important component that offers natural attractions—sunshine, sightseeing, and sporting facilities (Singhal 2006). Further, accommodation is one of the core areas of the tourism sector, which has an important contribution to the development of the tourism industry (Bhatia 2003).

Boniface and Cooper (2001) have mentioned the similar geographical components of tourism as Robinson (1976) described. They subdivided geographical resources for tourism as physical and cultural and further stated three main characteristics of geographical components of tourism. These characteristics are tangible objects of economic value. Tourism resources are shared with agriculture, forestry, water management, landscape, and accommodation.

Table 2.1 shows the geographical components of tourism (Robinson 1976). It is broadly divided into six sections—accessibility and location, space, scenery, animal life, settlement features, and culture. Further, the last four sections are divided into several sub-sections. The scenery has landforms (river valleys, middle altitudes, highlands, alpine pasturelands, and snow-clad Himalaya) water (rivers, lakes, waterfalls, geysers, and glaciers), climate (sunshine and clouds, temperature conditions, rain, and snow), and forest (national parks) as subcomponents. Animal life has wildlife—birds, game reservations, and zoos, and hunting and fishing as subcomponents. Settlement features have towns, cities, villages, historical remains and monuments,

V. P. Sati, *Sustainable Tourism Development in the Himalaya:
Constraints and Prospects*, Environmental Science and Engineering,
https://doi.org/10.1007/978-3-030-58854-0_2

Table 2.1 Geographical components of tourism (after Robinson 1976)

Accessibility and location	
Space	
Scenery	Landforms—river valleys, middle altitudes, highlands, alpine pasturelands, and snow-clad Himalaya
	Water—rivers, lakes, waterfalls, geysers, and glaciers
	Climate—sunshine and clouds, temperature conditions, rain, and snow
	Forests—national parks
Animal life	Wildlife—birds, game reservations, zoos
	Hunting and fishing
Settlement features	Towns, cities, and villages
	Historical remains and monuments Archaeological remains
Culture	Ways of life, traditions, folklore, art, and crafts

and archaeological remains as subcomponents. Culture describes the way of life, traditions, folklore—folk music and folk songs, art, and craft.

This chapter deals with geographical components of tourism development in the Uttarakhand Himalaya. As Uttarakhand is a mountainous state, therefore, the main components are related to the mountain features, which include mountain landscapes, river topographies, forest landscape, and feasible and varied climate. Data on different geographical components of tourism have been gathered from the Survey of India's toposheets and personal observations of the author. A map is digitalized, which shows the geographical basis of tourism. Other figures are developed by the author and photos are presented to enrich the chapter's contents.

2.2 Geographical Components of Tourism

Geographical components such as spectacular landscapes, river topographies, forest landscapes, and varied climate determine tourism development in the Uttarakhand Himalaya. Here, all these components/factors are plenty and unique. The author has divided these components into several subcomponents, which are shown in Fig. 2.1, and has presented their details in the following paragraphs.

Figure 2.2 shows the major geographical components of tourism such as major rivers, glaciers, mountain peaks, and cities and towns. There are numerous rivers and their tributaries in the Uttarakhand Himalaya out of which only major rivers are presented on the map. Similarly, the major cities/towns, mountain peaks, and glaciers are shown.

Fig. 2.1   Geographical components of tourism in the Uttarakhand Himalaya

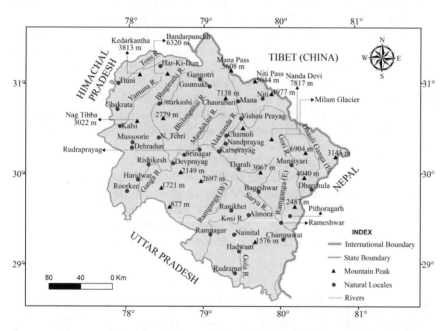

Fig. 2.2   Major geographical components of tourism—rivers, natural locals, glaciers, and mountain peaks

2.3   Spectacular Landscapes

The Uttarakhand Himalaya has spectacular landscapes along the altitudinal gradients ranging from the Doon valley to the Shivalik ranges, the Middle Himalaya, the Upper Himalaya, the alpine pasturelands, and the snow-clad mountain peaks. The Doon Valley stretches between the Ganga River in the east and the Yamuna River in the west. Mountain ranges surround it from north and south. Further, it is surrounded by dense monsoon forests of the RJNP. It is a fertile valley, extending between 29°57′30″E–30°30′30″N and 77°35′E–78°19′E. The length of the valley is 70 km and the width is 20 km (1400 km^2). The valley is mainly urbanized as about 156.7 km^2 area is under urban land. The landscape of Doon Valley is spectacular and the climate is feasible. It was known for the aromatic Basmati Rice during the recent past. The famous Doon School and the Army cantonment are located here. Besides, there are many national-level institutions. The beautiful landscape attracts a large number of tourists. Tarai and Bhabar are the low-lying areas of Nainital and USN districts. Tarai is a wet region while Bhabar is a dry and stony-forested region. Tourists visit Tarai and Bhabar during the winter as the climate remains very feasible around that time.

Shivalik ranges lie in the foothills of the Middle Himalaya, formed by the deposition of debris eroded from the Greater Himalaya. The Shivalik hills are covered by dense monsoon deciduous forests, with mammals and birds present in high diversity. Both the Garhwal and Kumaon Himalayas have Shivalik ranges, winter being the best season to visit them. The Middle Himalaya has the best tourist locations, where India's best hill resorts are situated. Figure 2.3 shows the beautiful view of Gopeshwar town (a) and Chakrata town (b), which are located in the Middle Himalaya. Gopeshwar town is located at an altitude of 1550 m, whereas Chakrata town is located at an altitude of 2200 m. Some of the famous hill resorts are Mussoorie, Nainital, Almora, and Ranikhet. Climatic conditions are very suitable during the summer season, and snowfalls in the upper-middle altitudes during the winter. Tourist visits the Middle Himalaya both in winter and summer.

The highlands and the alpine meadows are famously known for the trekking routes because many of the natural locales are situated above 3000 m. Most of them do not have appropriate transportation facilities therefore, the tourists visit these places by trekking all the way. Some of the famous trekking routes are the Valley of Flowers, Dayara Bugyal, Bedni Bugyal, Ali Bugyal, Roopkund, Pindari Glacier, Milan Glacier, and Gaumukh. The Great Himalayan Ranges are the major

(a)   (b)

Fig. 2.3 a Gopeshwar town, capital of Chamoli district. **b** Chakrata town in Dehradun district; *Photo* By author

tourist attraction. Mountaineering in the major mountain peaks such as Nanda Devi, Kamet, Trishul, Chaukhamba, Swargarohini, Nandaghunti, and Bhagirathi group of peaks is famous for adventure tourism. Figure 2.4 shows alpine pastureland before the Kedarnath pilgrimage (a) and the Mount Trishul in the Greater Himalaya (b).

Table 2.2 shows the mountain peaks of the Uttarakhand Himalaya. A total of 15 mountain peaks are presented where mountaineering is practiced. The Highest peak is Nanda Devi, which has 7817 m altitude. It is followed by Kamet (7756 m), Abi Gamin (7355 m), Chaukhamba (7138 m), and Trishul (7120 m). The lowest peaks

(a)   (b)

Fig. 2.4 a Alpine grasslands in Kedarnath pilgrimage. **b** The snow-clad Mount Trishul facing the Talwadi village; *Photo* By author

Table 2.2 Mountain peaks of the Uttarakhand Himalaya

S. No.	Name of the mountain peak	Altitude (m)
	Swargarohini	6252
	Bandarpooch	6316
	Shivling	6543
	Nilkantha	6596
	Meru Peak	6660
	Haathi Parvat	6727
	Nanda Kot	6861
	Panchchuli	6904
	Kedarnath	6940
	Dunagiri	7066
	Trishul	7120
	Chaukhamba	7138
	Abi Gamin	7355
	Kamet	7756
	Nanda Devi	7817

Source Survey of India's toposheets

are Swargarohini (6252 m) and Bandarpooch (6316 m). All these peaks are major attractions for tourists. The adventurers practice mountaineering on these mountain peaks.

Glaciers of the Garhwal and Kumaon Himalayas are trekked by the adventurers. The nine major glaciers in the Garhwal Himalaya and six major glaciers in the Kumaon Himalaya and their altitudes are presented (Table 2.3). These glaciers are the major sources of rivers of the Uttarakhand Himalaya. The main glaciers in the Garhwal Himalaya are Dunagiri (5150 m), Bandarpooch (4442 m), Gangotri (3892 m), Satopanth and Bhagirathi Kharak (3820 m), Dokriani (3800 m), Chaurabari Bamak (3800 m), Khatling (3717 m), Tipra Bamak (3600 m), and Nanda Devi group of glaciers (3300 m). The six major glaciers of the Kumaon Himalaya are Pindari (4625 m), Sunderdhunga (4320 m), Milam (4242 m), Ralam (4000 m), Kaphini (3840 m) and Namik (3600 m). These glaciers are the major attractions for adventurers and are considered as major trekking routes.

Table 2.3 Glaciers of Garhwal and Kumaon Himalaya

S. No.	The glaciers of the Garhwal Himalaya	Altitude (m)
1	Nanda Devi group of glaciers	3300
2	Tipra Bamak glacier	3600
3	Khatling glacier	3717
4	Chaurabari Bamak	3800
5	Dokriani glacier	3800
6	Satopanth and Bhagirathi Kharak glacier	3820
7	Gangotri glacier	3892
8	Bandarpooch glacier	4442
9	Dunagiri glacier	5150
The glaciers of the Kumaon Himalaya		
1	Namik glacier	3600
2	Kaphini glacier	3840
3	Ralam glacier	4000
4	Milam glacier	4242
5	Sunderdhunga glacier	4320
6	Pindari glacier	4625

Source Survey of India's toposheets

2.4 River Topographies

The rivers of the Uttarakhand Himalaya are not just physical features but also the centers of cultural beliefs. The topographies of these rivers are panoramic. Further, the cities/towns, situated on the banks of these rivers, are the major pilgrimages of the Uttarakhand Himalaya, visited by hundreds of thousands of pilgrims every year. Kedarnath is situated on the bank of the Mandakini River, Badrinath is situated on the bank of the Alaknanda River (also known as Vishnu Ganga), Gangotri is situated on the bank of the Bhagirathi River, and Yamunotri is situated on the bank of the Yamuna River. These four towns are the highland pilgrimages. Rishikesh and Haridwar, the river valley pilgrimages, are situated on the bank of the Ganga River. Along with these pilgrimages, nine natural and cultural places are situated on the confluence of the two rivers are known as the Prayags. On the Alaknanda River, there are six Prayags—Keshav Prayag on the Alaknanda and the Saraswati Rivers, Vishnuprayag on the Alaknanda and the Dhauli Ganga, Nandprayag on the Alaknanda and the Nandakini Rivers, Karnprayag on the Alaknanda and the Pindar Rivers, Rudraprayag on the Alaknanda and the Mandakini Rivers, and Devprayag on the Alaknanda and the Bhagirathi Rivers. Son Prayag is situated on the meeting point of the Mandakini River and the Son Ganga. Bageshwar is located on the banks of Gomati and the Saryu Rivers. One of the Prayags is called Suryaprayag, situated near Ghansali town on the banks of the Bal Ganga and Dharma Ganga, tributaries of the Bhilangana River (Sati 2019a). Figure 2.5 shows (a) the landscape of the Alaknanda River flowing in the Srinagar Garhwal valley and (b) the Ganga River is making a big curve below Sakinidhar before Vyasi village.

There are three major river systems in the Uttarakhand Himalaya (Table 2.4). The first one is the Yamuna system which has Tons, Rupin and Supin, and Pawar River as tributaries. The Yamuna River originates from the Yamunotri glacier in Uttarkashi district and flows through Dehradun district. The Ganga system in Uttarakhand is the biggest one which has three major sub-systems—the Bhagirathi, the Alaknanda,

(a)   (b)

Fig. 2.5 a The Alaknanda river flowing near Srinagar town. **b** The Ganga river flowing down to Sakinidhar; *Photo* By author

Table 2.4 The major river systems and their tributaries

S. No.	Major river systems	Tributaries
1	The Yamuna system	The Yamuna River The Tons River Rupin and Supin Rivers The Pawar River
2	The Ganga system (1) The Bhagirathi sub-system	The Bhagirathi River The Asi Ganga The Bhilangana River The Bal Ganga The Dharma Ganga
	(2) The Alaknanda sub-system	The Alaknanda River The Saraswati River The Dhauli Ganga The Nandakini River The Pindar River The Mandakini River The Madhyamaheshwar River The Kali Ganga The Son Ganga
	(1) The Ramganga sub-system	The Ramganga River (W) The Kosi River
3	The Kali system	The Kali/Sharda River The Ramganga River (E) The Gori River The Dhauli River (E) The Gomati River The Saryu River

and the Ramganga (W). The Bhagirathi sub-system comprises of the Bhagirathi River, the Asi Ganga, the Bhilangana River, the Bal Ganga, and the Dharma Ganga. The Alaknanda sub-system comprises of the Alaknanda River, The Saraswati River, The Dhauli Ganga, the Nandakini River, the Pindar River, and the Mandakini River along with their numerous tributaries. The Ramganga sub-system comprises of the Ramganga River (W) and the Kosi River. The third river system is the Kali System. The main rivers in the Kali River system are the Kali/Sharda River, the Ramganga River (E) the Gori River, The Dhauli River (E), the Gomati River, and the Saryu River. They all are perennial and very important rivers. These rivers make beautiful landscapes from their origin to the confluence. They make a V-shaped valley, gorges, and waterfalls. Many tourist/pilgrim centers (towns and villages) are located along the river banks. The rivers have spiritual importance in the Uttarakhand Himalaya. The pilgrims worship the rivers during their visit to Uttarakhand. River rafting (adventure tourism) has been developed during the recent past as mass tourism. The major rivers where river rafting is practiced are the Alaknanda, the Yamuna, the Ramganga, and the Kali rivers.

2.5 Forest Landscapes

The forests of the Uttarakhand Himalaya make beautiful landscapes. From the monsoon deciduous forests to scrubs, pine, mixed oak, coniferous, and the alpine grasslands, the forests have uniqueness (Sati 2019b). These forests are vertically and horizontally distributed. The monsoon deciduous forests are found up to 500 m altitudes. The bushes of the river valleys are found between 500 and 1000 m along with some other local species of plants. Pine is extensively found between 1000 and 1800 m, making a beautiful landscape. Mixed-oak forests are found from 1800 to 2200 m, and above it, coniferous forests including deodar, fir, spruce, and cedar are found. Tree-line can be seen up to 3000 m. Extensive alpine grasslands are found up to 3500 m, where flowers blossom during the autumn season. The famous Valley of Flowers is situated in alpine grasslands. These forests are home to many mammals and birds, where tourists, particularly the nature lovers, visit. Figure 2.6 shows the forest landscape—monsoon deciduous forest in RJNP (a) and coniferous deodar forest around Mussoorie town (b).

Uttarakhand has a total of 71.05% forest area. Because of rich biodiversity, 12% of its geographical area is protected, which includes 6 National Parks, 7 Wildlife Sanctuaries, 4 conservations, and 1 biosphere reserve. It has 102 mammals, 600 birds, 19 amphibians, 70 reptiles, and 124 species of fish (FSI 2017). The rich biodiversity, national parks, and wildlife sanctuaries provide a suitable base for tourism activities mainly ecotourism. The major parks and wildlife sanctuaries where ecotourism is practiced are RJNP, CWS, CNP, VOFNP, NDBR, GNP, GWS, and WNP.

Fig. 2.6 Forest landscape **a** Monsoon deciduous forest in RJNP. **b** Coniferous deodar forest around Mussoorie town; *Photo* By author

2.6   Feasible and Varied Climate

The climate of Himalaya varies from subtropical to temperate, cold, and frigid cold. The Middle Himalaya, the highlands, and the pasturelands have very feasible climatic conditions during the summer season. India's hill resorts are located in the Middle Himalaya, which are the major tourist destinations. The summer season is quite suitable for tourism in these resorts. *Yatra* season starts from April and ends in the last week of October. This season is also conducive for trekking and mountaineering. During the entire *Yatra* season, the climate remains conducive. Tourism is also practiced during the winter. Snowfalls in the Middle Himalaya and tourists visit here to enjoy it. For river rafting, the winter season is feasible.

2.7   Rural Landscape

About 70% of the population live in rural areas in the Uttarakhand Himalaya. Further, about 95% of villages are located in the mountainous mainland. Population concentration is the highest in the Middle Himalaya because of the landscape and availability of arable land. The accessibility to these villages is substantial. The landscape, climate, and environmental conditions of the villages are very conducive. Further, these villages have a rich culture, customs, and rituals. Both, beautiful landscapes and rich culture, serve as major attractions for tourists and major factors for rural tourism development. The rich traditional food and beverages, and welcoming behaviour of the local people also support rural tourism. Rural tourism can also be termed as village tourism, farm tourism, and peace tourism (Sati 2019c). In the Uttarakhand Himalaya, there are several places for rural tourism such as Munsiyari in Pithoragarh district, Chopta in Rudraprayag district, and Gwaldam in Chamoli district. Figure 2.7

(a)   (b)

Fig. 2.7   Panoramic view of the rural landscape **a** Village Kaub in the Pindar River watershed, **b** a village near Pauri town in Salan region; *Photo* By Author

shows a panoramic view of the rural landscape. A village named Kaub in the Pindar River basin (a) and a village situated in the Nayar River valley in Pauri district (b) are shown (Fig. 2.7).

2.8 Conclusions

Geographical components of tourism are described. It has been noticed that the Uttarakhand Himalaya has rich geographical attributes in the forms of natural landscapes, forest landscapes, river topographies, climate, and spectacular rural locales. These geographical attributes provide a suitable platform for tourism development. The need of the hours is to develop these destinations for sustainable tourism development. Adequate infrastructural facilities can be developed and provided to tourists. It has been observed that the natural beauty of many locales has not been harnessed optimally for tourism development because these locales have not been much publicized. Institutional support for the publicity of these locales is required so that the potential of geographical attributes can be harnessed and sustainable tourism development can be attained.

References

Bhatia AK (2003) Tourism development, principles and practices, Steerling Publishers Pvt. Ltd, New Delhi, p 4

Boniface B, Cooper C (2005) World destination casebook: the geography of travel and tourism. Elsevier, Amsterdam, p 268

FSI (2017) Forest Survey of India Report 2017, Dehradun

Robinson H (1976) A geography of tourism, Ply Mouth, Mac Donald and Events Ltd., p 40

Sati VP (2019a) Potentials and forms of sustainable village tourism in Mizoram, North East India. J Interdisc Acad Tour 4(1):49–62. https://doi.org/10.31822/jomat527278

Sati VP (2019b) Himalaya on the threshold of change. Springer International Publishers, Switzerland, p 250. https://doi.org/10.1007/978-3-030-14180-6

Sati VP (2019c) Forests of Uttarakhand: diversity, distribution, use pattern and conservation. ENVIS Bulletin Himalayan Ecol 26, 21–27

Singhal GD (2006) Glimpes of tourism in India, Kanishka Publishers, Distributors, New Delhi, p 171

Chapter 3
Cultural Components of Tourism Development

3.1 Introduction

Tourists travel outside their spaces to experience different environments and cultures (Selanniemi 2003; Shields 1991). Culture, as a resource, is a basis of tourism both domestic and international. Further, cultural diversity supports more tourist interactions and strengthens the local culture. It comprises both ways of life (beliefs, values, social practice, rituals, and traditions): tangible—buildings, monuments, objects; and intangible—language, performances and festivals, and craftsmanship, which are called cultural products. For tourism practices culture refers to people and their social characteristics, traditions, and customs. Each cultural product has its importance and values, which the tourists/pilgrims utilize while they travel. UNESCO has declared several cultural locales of the Uttarakhand Himalaya as cultural heritages/landscapes. These cultural heritages provide input to tourists/pilgrims to know more about them. A tourist/pilgrim interacts and gets wide experiences from a diverse culture. On the other hand, cultural diversity manifests more attraction of tourists within the area, leading to economic development. Cultural creativity is a dynamic process that is related to intellectual, technological, and institutional senses.

Cultural resources are the bases of economic development, which can be achieved through tourism practices. Further, tourism can be an instrument for income generation, employment, contributing to economic development. In this process, cultural resources have a significant contribution. Since the mid twentieth century, mobilization of tourists/pilgrims is a preferred form of economic development at all three levels—local, regional, and global (UNESCO 1999). Cultural tourism has prospects to enhance economic diversification (UNCTAD 2004). However, in many countries, cultural tourism could not be utilized substantially. Cultural tourism is an important instrument for exchange and dialogue. A successful existence of culture depends on its contact with other cultures and tourism can play an important role in contacting and

V. P. Sati, *Sustainable Tourism Development in the Himalaya:
Constraints and Prospects*, Environmental Science and Engineering,
https://doi.org/10.1007/978-3-030-58854-0_3

making intellectual dialogue (Viard 2000). Tourism also involves the strengthening of culture in various ways.

About 40% of the international tourists in 2016 were involved in cultural activities (UNWTO 2016). In recent times, culture has become a key product of tourism, both domestic and foreign. Simultaneously, cultural space is a major attraction of cultural tourism (OECD 2009; ETC 2005). Richards (2018) stated that culture has an important role in tourism development. Cultural tourism is mainly associated with understanding and experiencing the culture and customs of the area, the role of education in this regard being noteworthy. It is mainly engaged with art, crafts, festivals, foods, and folklore—dances and songs. In Himalaya, cultural tourism is being practiced from time immemorial. Here, cultural tourism is viewed as an important, potential source of tourism growth. It is not only considered as the global industry to bring income but also a national identity and a means for protecting culture and cultural heritage (Richards 2007). It also serves as an alternative to mass tourism (Edgell 2006).

The Uttarakhand Himalaya has been the center of pilgrimage tourism and spirituality for time immemorial. It practices cultural, religious, and spiritual tourism, and known for its history, culture, beliefs, and traditions, which are widely admired and celebrated. It has a unique cultural landscape, wildlife, and people. The two cultural landscapes of the Uttarakhand Himalaya—Garhwal and Kumaon have their uniqueness in tradition and beliefs. Social stratification is also unique, as Brahmins, Rajputs, Harijans, and tribes comprise the society and live with harmony in different parts of the state under diverse geo-environmental and socio-cultural conditions.

The belief of '*Atithi Devo Bhavah*' (the Guest is God) is very popular. The folks are humble, simple, honest, and are of welcoming nature. They are referred to as *Paharis* (hill people). Diversity in races and ethnicity is high. The majority of people are involved in practicing agriculture and allied works, artisans, and making handicrafts. They are nature lovers and worship all components of nature.

Their clothing style and types are different as compared with that of people in the other parts of the country. Clothing represents the culture and customs of the region. Apart from the *Pahari* community, Punjabis, Tibetan, and Nepalese have their styles of dresses. The *Pahari* people utilize local resources for making their clothes such as wool obtained from sheep, using which they make sweaters and jackets. Women wear a cloth called *Pakhula* and men wear *Kurta* and *Pajama*. Men also wear Dhoti. *Ghagra, Lehenga*, and *Choli* are traditional dresses, which women wear. Art and crafts of the Uttarakhand Himalaya are famous worldwide. Artisans carry out making art and crafts as their full-time work. Here, some awe-inspiring and astounding artworks, made by using wood, jute, and hemp, can be seen, which reflect the culture and customs of the region. Some of the important art and crafts are embroidered carpets, cushion covers, bedsheets, and woolen knitted wear. Lifestyle varies, both in urban and rural areas, with high heterogeneity and distinctness. The people of rural areas are comparatively poor and less educated therefore, their lifestyle is different, while the urban areas are more advanced. In this chapter, various cultural components of tourism are elaborated. The culture of the region is described in terms of tourism development. The pilgrimages—the highlands and the river valleys, fairs and festivals, traditional foods and beverages, and handmade art and crafts, which are the major tourism components are described. The contents of this chapter are based

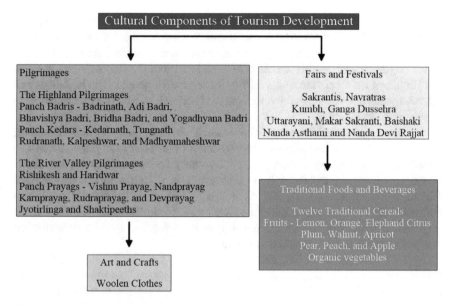

Fig. 3.1   Cultural components of tourism

on the in-depth and wide observation and experience of the author on the culture of the Uttarakhand Himalaya, as he has traveled the entire region several times for geographical excursion.

3.2   Cultural Components of Tourism

Cultural components of tourism in the Uttarakhand Himalaya include the pilgrimages—the highlands and the river valleys. They are further divided as Panch Badri, Panch Kedar, and Panch Prayag. Shaktipeeths and Jyotirlingas are the centers of tourism attraction. Fairs and festivals, traditional foods and beverages, and art and crafts also support tourism, being the major tourist attractions. All these components of tourism are centuries old, referred to in the Hindu religious pearls of wisdom such as Ramayana and Mahabharata. For example, it is believed that the temple of Kedarnath was built by the Pandavas. Figure 3.1 shows the major cultural components of tourism in the Uttarakhand Himalaya.

3.3   Pilgrimages to the Himalaya

Pilgrimages to the Himalaya are century-old practices. They have been an important part of Hindu culture and tradition from time immemorial (Bharati 1978; Hausner

2007; Singh 2013). The whole Himalayan region has an important spiritual meaning
for Hindus. They believe it to be a 'sacral space' (Grotzbach 1994). Worshipping of
the '*Char Dhams*' (four pilgrimages) , holy rivers, and nature's deities has its roots
in the Aryan culture, which was later on integrated into Hinduism. Pilgrimages to
the Himalaya have become a way of life for the native people. The highlands and
the river valleys have deep-rooted rich cultural heritage, which can be visualized in
these pilgrimage centers.

The four highland pilgrimages—Badrinath, Kedarnath, Gangotri, and Yamunotri,
and the two river valley pilgrimages—Rishikesh and Haridwar provide bases to
cultural tourism although, there are numerous other pilgrimages. Tourism data shows
that out of the total 434.03 million tourists/pilgrims' inflow in the Uttarakhand
Himalaya during 2000–2018, 300.48 million pilgrims visited the highlands and
the river valleys pilgrimages, which accounts 69.23%. In Haridwar, 240.32 million
pilgrims visited during the period, which was 55.37% of the total tourists/pilgrims
that have arrived in the Uttarakhand Himalaya.

Figure 3.2 shows the major cultural places of the Uttarakhand Himalaya. It shows
that these places are located mainly in the Garhwal Himalaya. The Kumaon Himalaya
has few cultural places. In the Garhwal Himalaya, the highest number of cultural
places is located in the Chamoli district. Rudraprayag, Tehri, and Uttarkashi districts
have an equal number of cultural places. Haridwar and Dehradun districts have
one major cultural place i.e. Haridwar and Rishikesh, respectively. Cultural places,
located in Chamoli and Rudraprayag districts, such as Panch Badri—Badrinath,

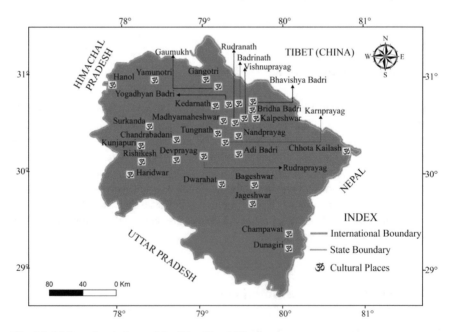

Fig. 3.2 Major cultural places of the Uttarakhand Himalaya

Yogadhyan Badri, Bridha Badri, Bhavishya Badri, and Adi Badri; Panch Kedar—
Kedarnath, Madhyamaheshwar, Rudranath, Tungnath, and Kalpeshwar; and Panch
Prayag—Vishnuprayag, Nandprayag, Karnprayag, Rudraprayag, and Devprayag are
prominent. Panch Badri and Panch Kedar are closely located to each other. Gaumukh,
Gangotri, and Yamunotri are the major cultural places of Uttarkashi district. Hanol is
an important cultural place located in the Dehradun district. Rishikesh pilgrimage is
sprawled in three districts—Dehradun, Tehri, and Pauri. The famous Shaktipeeths—
Kunjapuri, Chandrabadni, and Surkanda are located in Tehri district. One of the
Jyotirlingas is Joshimath, located in the Chamoli district. Devprayag is a famous
cultural place, located in Tehri district on the confluence of the rivers—Bhagirathi and
the Alaknanda in Tehri district. After confluence at Devprayag, these two rivers are
called 'the Ganga'. Bageshwar town is located in Bageshwar district, where 20 days
Magh Mela is celebrated. Jageshwar having a group of more than 80 temples, lies on
the bank of the Jataganga, in a picturesque and serene environment. Dwarahat, known
as 'Old Dwarika' lies in Almora district. Champawat and Dunagiri in Champawat
district and Chhota Kailash (Adi Kailash) in Pithoragarh district are the famous
cultural places of the Uttarakhand Himalaya.

3.4 Fairs and Festivals

Uttarakhand is an abode of folk deities. Every month, one or more fairs and festivals
are celebrated. The first day of every month, according to the Hindu calendar, is called
Sakranti. Baishakhi, celebrated from the first day of Baishakh month, is a week-long
celebration. The folk deities take a dip in the Ganga River and its tributaries, and
people worship them. Harela is celebrated on the first day of Shravan. Trees are
planted in and around the villages and agricultural fields. Ghee *Sakranti* is celebrated
on the first day of Bhado (Bhadrapada). It is related with Ghee, the milk product. The
production of milk is the highest during Bhado because of the high availability of
green fodder therefore people celebrate it as Ghee *Sakranti*. Besides, *Uttrayani, Magh
Mela, Fooldehi, Makar Sakranti*, and *Basant Panchami* are celebrated. *Navratras* are
celebrated two times in a year—during the months of Chaitra and Ashwini (Shardey).
Figure 3.3 shows a Doli (statue) of Lord Shiva (a) and people worshiping Lord Shiva
during the Baishakhi festival in April in the Pindar River basin (b).

The folks worship trees, water, land, and food. They believe that the almighty
lives everywhere. Figure 3.4 shows the cultural landscape. The bells are hanging on
the branches of a peepal tree in a village named Parethi (b), a temple in the dense
forest area in the same village (b).

The Kumbh Mela is celebrated in the Har-Ki-Pauri in Haridwar (Fig. 3.5a). It is
celebrated once in every 12 years. Half Kumbh Mela is also celebrated once in every
six years. It is a month-long fair. Hundreds of thousands of pilgrims, from within
and outside the country, visit during the Kumbh fair. The pilgrims believe that taking
a dip in the Ganga River in Har-Ki-Pauri during the Kumbh fair helps them get rid

Fig. 3.3 **a** a Doli of Lord Shiva (folk deity), **b** devotees celebrating Baishakhi fair along with the folk deity; *Photo* By author

Fig. 3.4 Cultural landscape **a** folks worship tree, **b** a temple of forest deity; both are in the middle catchment of the Pindar River; *Photo* By author

Fig. 3.5 **a** Kumbh Mela is being celebrated at Har-Ki-Pauri, Haridwar, **b** NDRJ Yatra is on its way to Roopkund; *Photo* By author

of the cycle of birth and death and in achieving immortality. Kumbh means a pot of nectar, immortality.

Every year, many spiritual processions (*Yatras*) take place. The procession of folk deities along with the devotees proceeds from one village to another for a certain period. NDRJ *Yatra* is the longest one, which starts from the Nauti village of Karnprayag administrative sub-division and ends in Roopkund, a highland mysterious lake, situated on the lap of Mount Trishul at an altitude of 4800 m. With a 280 km route (both ways), passing through several villages, it takes 20 days to complete the procession (Sati 2017). The landscape of the route is spectacular. It goes through the river valleys, the Middle Himalaya, the Alpine meadows, and the snow-clad mountain peaks. The Doli of Goddess Nanda leads the procession. The Goddess Nanda is the adorable deity. It is believed that the Goddess Nanda, born in Nauti village, married Lord Shiva, who lives in the Himalaya. Every year, Goddess Nanda is sent to Kailash in the month of Bhadrapada. However, the main procession takes place once every 12 years. Hundreds of thousands of devotees take part in the procession (Fig. 3.5b). Kailash-Mansarovar *Yatra* is performed between June and October every year. It goes through the Pithoragarh district.

Ganga Dussehra is celebrated on the 10th day of Jyestha month (the last week of May) according to the Hindu calendar. It is celebrated to mark Gangavataran, the descent of the Ganga on the Earth from Heaven. It is celebrated all along the Ganga River, from Gaumukh to Haridwar mainly in Gangotri, Uttarkashi, Devprayag, and Haridwar towns/cities. A huge number of pilgrims, from within and outside the country, visit these river valley cultural locales and take a dip on the Ganga River on this auspicious occasion. The Ganga's water is pure and the pilgrims believe that taking a dip in the Ganga helps them get rid of the cycle of birth and death.

3.5 Traditional Foods, Art, and Crafts

The Uttarakhand Himalaya has rich traditional food varieties, handmade arts, and crafts. The traditional food is served to tourists and pilgrims during the Yatra season. Handcrafts are made of wool and bamboo. It produces sheep wool, which is used for making several woolen cloths. Bamboo is used for making various kinds of articles. Further, the entire region produces various kinds of fruits—mango types, citrus, nut, and apple. All these drivers/items support cultural tourism development in the Uttarakhand Himalaya.

3.6 Conclusions

The description of the cultural components of tourism reveals that the Uttarakhand Himalaya has a rich and unique culture and cultural heritage for sustainable tourism development. The pilgrimages are the major destinations for cultural, spiritual, and religious tourism. It has been noticed that pilgrims, who visit Uttarakhand outnumber the tourists. The rich culture, customs, fairs, and festivals of the Uttarakhand Himalaya are the major attractions of tourism. Adequate infrastructural facilities—transportation, accommodation, and institutional support can be developed in these cultural locales to boost up the sustainable tourism development.

References

Bharati A (1978) Actual and ideal Himalayas: Hindu Views of the mountains. In: Fisher J (ed) Himalayan anthropology: The Indo-Tibetan interface. Mouton, Hague, pp 77–82
Edgell DL (2006) Managing sustainable tourism: a legacy for the future. Haworth Hospitality Press, New York
ETC (2005) City tourism and culture. European Travel Commission, Brussels
Grotzbach E (1994) Hindu-Heiligtümer als Pilgerziele im Hochhimalaya. In: Erdkunde 48(3):48
Hausner S (2007) Wandering with Sadhus: Ascetics in the Hindu Himalayas. Indiana University Press, Bloomington
OECD (2009) The impact of culture on tourism. OECD, Paris
Richards G (2007) Cultural tourism: global and local perspectives. Haworth Press, NY
Richards G (2018) Cultural tourism: a review of recent research and trends. J Hospital Tour Manage 36. https://doi.org/10.1016/j.jhtm.2018.03.005
Sati VP (2017) Cultural geography of Uttarakhand Himalaya. Today and Tomorrow, Printers and Publishers, Delhi
Selanniemi T (2003) On holiday in the Liminoid playground: play, time, and self in tourism. In: Bauer TG, McKercher B (eds) Sex and tourism: journeys of romance, love, and lust. Haworth, New York/London/Oxford, pp 19–34
Shields R (1991) Places on the margin—alternative geographies of modernity. Routledge, London
Singh RPB (2013) Pilgrimage-tourism: perspective & vision. In: Hindu tradition of pilgrimage: sacred space and system. Dev Publishers, New Delhi, pp 305–332. ISBN (13): 978-93-81406-25-0
UNCTAD (2004) World investment report 2004. In: The shift towards services. United Nations Conference on Trade and Development, Geneva
UNESCO (1999) Towards new strategies for culture in sustainable development. Paper jointly presented at the 1999 international conference 'culture counts', in Florence, Italy. UNESCO Publishing, Paris
UNWTO (2016) World tourism organization annual report 2016. United Nations World Tourism Organization, Madrid
Viard J (2000) Court traité sur les vacances, les voyages et l'hospitalité des lieux. Aube, Paris

Chapter 4
Types of Tourism and Major Tourists/Pilgrims' Centres

4.1 Introduction

The Uttarakhand Himalaya is endowed with plenty of world-famous natural locales, the highlands, and the river valleys pilgrimages, and spectacular landscapes from the river valleys to middle altitudes, highlands, alpine pasturelands, and the perpetual snow-clad mountain peaks. The major types of tourism—natural tourism, pilgrimage/cultural tourism, park and wildlife tourism (eco-tourism), and adventure tourism are practiced here (Government of India 2008; Kohli 2002; Nakajo 2017; Nakatani 2011; Sharma 2008; Siddiqui 2000; Sati 2013, 2015). Uttarakhand has several national parks and wildlife sanctuaries among them NDBR, RJNP, CNP, GWS, GNP, BWS, and Asan bird sanctuary are famous. Many tourists visit these parks and sanctuaries every year. Trekking and mountaineering are also practiced however, few tourists perform them. The major trekking routes are the Valley of Flowers, Roopkund, Bedni Bugyal, Pindari glacier, Dayara Bugyal, Gaumukh, Yamunotri, Kedarnath, and Milam glacier. Mountaineering to Nanda Devi, Chaukhamba, Bhagirathi peak, Trishul, Kamet, Nandaghunti, Dunagiri, Purnagiri, and Dronagiri is very adventurous. River rafting is very popular as there are many famous rivers, which are ideal for river rafting such as the Alaknanda River, the Tons River, the Ramganga River, and the Kali River. In the meantime, many suitable camping sites are available in these river banks. Auli in Chamoli district has been developed as a major winter skiing center. Badrinath, Kedarnath, Gangotri, and Yamunotri are the major highland pilgrimages, and Rishikesh and Haridwar are the major river valleys pilgrimages. Besides, there are Panch Badri, Panch Kedar, and Panch Prayag. Mussoorie, Chakrata, Almora, Ranikhet, Nainital, Pithoragarh, and Munsiyari are the major hill resorts. In this chapter, all types of tourisms practiced in the Uttarakhand Himalaya are described. Similarly, a brief description of some of the important tourist/pilgrim centers, are described. A map is digitalized showing natural and cultural locales, and other figures are developed by the author. Most of

© The Editor(s) (if applicable) and The Author(s), under exclusive license to Springer Nature Switzerland AG 2020
V. P. Sati, *Sustainable Tourism Development in the Himalaya: Constraints and Prospects*, Environmental Science and Engineering, https://doi.org/10.1007/978-3-030-58854-0_4

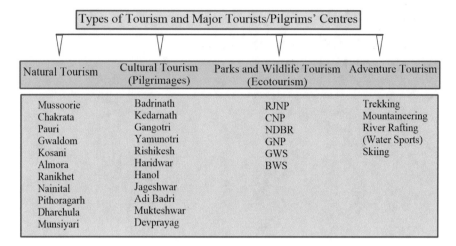

Types of Tourism and Major Tourists/Pilgrims' Centres			
Natural Tourism	Cultural Tourism (Pilgrimages)	Parks and Wildlife Tourism (Ecotourism)	Adventure Tourism
Mussoorie	Badrinath	RJNP	Trekking
Chakrata	Kedarnath	CNP	Mountaineering
Pauri	Gangotri	NDBR	River Rafting
Gwaldom	Yamunotri	GNP	(Water Sports)
Kosani	Rishikesh	GWS	Skiing
Almora	Haridwar	BWS	
Ranikhet	Hanol		
Nainital	Jageshwar		
Pithoragarh	Adi Badri		
Dharchula	Mukteshwar		
Munsiyari	Devprayag		

Fig. 4.1 Types of tourism and major tourists/pilgrims' centers

the description in this chapter is based on the observations of the author about the region.

4.2 Types of Tourism and Major Tourists/Pilgrims' Centres

The Uttarakhand Himalaya practices four types of tourism in the numerous tourists/pilgrims centers. The main tourism types practiced here are natural tourism, cultural tourism (pilgrimages), park and wildlife tourism (ecotourism), and adventure tourism. Besides, rural tourism and health tourism are also suitable to be practiced. Figure 4.1 shows the types of tourism and major tourists/pilgrims' centers. Major towns and cities in natural tourism, pilgrimages in cultural tourism, parks, and sanctuaries in park and wildlife tourism, and major treks and sports places in adventure tourism are described.

Figure 4.2 shows a map depicting the major tourist places and pilgrimages. The Garhwal Himalaya has more number of natural and cultural locales than the Kumaon Himalaya. A detailed description of the tourist places shown on the map is presented in the following paragraphs.

4.3 Natural Tourism

The Uttarakhand Himalaya has numerous natural locales and hill resorts, where an exodus number of tourists visit every year. Most of the hill resorts are located in the Middle Himalaya where the climate is very feasible mainly during the summer

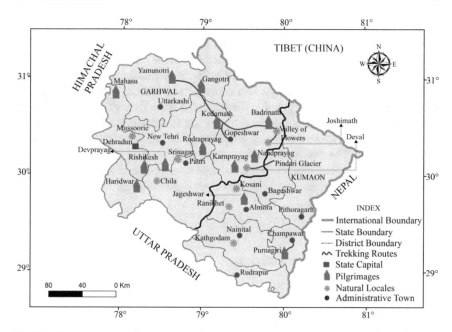

Fig. 4.2 Tourism types and the major tourist/pilgrim centers

season. In the meantime, the mainland of India, mainly the Ganges valley, the Thar (Rajasthan) Desert, and the Deccan plateau receive sunstrokes and heatwaves. To escape from the sunstrokes, the people of the plain regions come to Uttarakhand for leisure tourism in the hill resorts. The tourists also visit during winter when these Middle Himalayan hill resorts receive heavy snowfall. Although, there are many hill resorts and natural locales for tourism yet some of the important resorts are Mussoorie, Chakrata, Pauri, Gwaldam, Kausani, Almora, Ranikhet, Nainital, Pithoragarh, Dharchula, and Munsiyari. A detailed description of these natural locales and hill resorts, which are major tourist/pilgrim circuits and tourist routes, respectively, are given in Chaps. 6 and 7.

4.4 Cultural Tourism (Pilgrimages)

The Uttarakhand Himalaya has a rich culture and cultural heritage. It is called the land of the 'Gods and Goddesses'. The folks worship water, forests, animals, and the mountains—the life-supporting layers. The sun and the moon are also worshiped on no moon and full moon days. Food and water are believed to be the forms of Gods and Goddesses. Fairs and festivals are celebrated throughout the year with the change in seasons. Both male and female deities are worshiped. The main deities are Lord Shiva and the Goddess Shakti, and their numerous incarnations/forms. The entire Uttarakhand has thousands of temples. NDRJ, Kumbh, Baishakhi, Magh Mela, Uttarayani, Ganga Dussehra, Sakrantis, and Navratras are the major folk fairs and festivals. Hundreds of thousands of pilgrims celebrate these fairs and festivals.

The Uttarakhand Himalaya is known for its highlands and the river valleys pilgrimages. Pilgrimage to the Himalaya is a century-old practice. Besides Hindus, the followers of Jainism, Buddhism, and Sikhism visit these river valleys and the highlands pilgrimages. The main pilgrimage sites are Badrinath, Kedarnath, Gangotri, Yamunotri, Hanol, Jageshwar, Panch Badri, and Panch Kedar. These are the highland pilgrimages. The river valleys pilgrimages are Rishikesh, Haridwar, and Panch Prayag. A detailed description of highland and river valleys pilgrimages are given in Chaps. 6 and 7, which are major tourist/pilgrim circuits and tourist routes, respectively.

4.5 Park and Wildlife Tourism

Park and Wildlife tourism mean watching wild and undomesticated animals in their normal home. Uttarakhand is bestowed with stunning national parks and adventurous wildlife sanctuaries. The tourists—nature lovers and eco-tourists visit these parks around the year for leisure tourism and holiday home. Rich faunal and floral diversity is found in these parks. NDBR is known for endangered musk dears, CNP is famous for tiger safari, RJNP is well known for elephant safari, and Asan barrage is a home for migratory birds. Most of the endangered species of fauna and flora are found in the parks and wildlife sanctuaries. These parks and sanctuaries are big in area and their landscape is spectacular. CNP and RJNP have mainly plain topographies. However, GNP, GWS, BWS, and NDBR are mountainous. Asan barrage also lies in the plain region on the Asan stream, a tributary of the Yamuna River. Forest types in these parks and sanctuaries vary from monsoon deciduous (CNP and RJNP) to pine, mixed-oak, and coniferous. The economic value of these forests is substantial. The major wildlife species found here are the Indian tiger, black bear, sloth bear, jungle cat, deer, leopard, languor, common otter, rhesus monkey, Indian elephant, ghoral (goat antelopes), porcupine, wild dog, king cobra, python, and a variety of migratory and resident birds. Uttarakhand has 12 Wildlife Sanctuaries and National Parks, stretching in around 13.8% of the total geographical area of the state. The two conservation reserves—Jhilmil Tal and the Asan Barrage and the two world Heritage sites—the NDBR and the VFNP are important sites for the region.

CNP in Nainital district, RJNP in the Doon Valley, GWS and GNP in Uttarkashi district, VFNP in Chamoli district, NDBR and Kedarnath Sanctuary in Rudraprayag and Chamoli districts, BWS and Askot Musk Deer Sanctuary in Almora district, Nandhaur Wildlife Sanctuary and Jhilmil Conservation Reserve in Nainital district, Binog Wildlife Sanctuary in Mussoorie (Dehradun district), Asan Barrage in Dehradun, and Sonanadi Wildlife Sanctuary in Nainital are famous national parks, wildlife and bird sanctuaries.

4.6 Adventure Tourism

Adventure tourism includes trekking, river rafting, mountaineering, and skiing that is practiced in the Uttarakhand Himalaya. In this chapter, adventure tourism and its types are described precisely. Trekking is one of the most important types of adventure tourism practiced in the Uttarakhand Himalaya. In the following paragraphs, the major trekking routes of the Garhwal and the Kumaon Himalayas are described.

4.7 Major Trekking Routes of the Garhwal Himalaya

4.7.1 Kedartal Trek

The trek has 5000 m altitude and it is a seven days long trek with a moderate difficulty level. The suitable months are May–June and September. The trek, located in the Gangotri area, Uttarkashi district, Garhwal Himalaya, has a spectacular landscape as it offers remarkable views of the Great Himalayan ranges—the Jogin peaks, the Bhirgupanth peak, and the Thalayasagar peak. Kedartal, also known as the Shiva' lake, is a hidden gem of the Himalaya, a 17 km trek from Gangotri. Kedarkharak is a camping site. Another important site is Bhojkharak near Kedar Ganga gorge. The main wildlife species found here are Blue Sheep, Goral, and Black Bear.

4.7.2 Rupin Pass Trek

Rupin pass trek lying in the Uttarkashi district has a maximum altitude of about 5000 m. It can trek within eight days and the best months to trek in Rupin pass are May–June and September–October. The trek is not much difficult. The Rupin pass is spectacular. It has a panoramic landscape, beautiful waterfalls, alpine pasturelands, and it is surrounded by lofty snow-clad mountain peaks. The pass provides a way to Himachal Pradesh from Uttarakhand. The trek starts from a township named Dhaula and goes up to the Rupin Pass. A beautiful village named Jakha lies on the way and the people of the village serve the trekkers by providing food and accommodation.

4.7.3 Kedarkantha Trek

It is comparatively an easy trek, which has a maximum altitude of 4000 m. About five days are needed to complete the trek and the ideal trekking months are from December to April. This trek, located in the GWS of Uttarkashi district, starts from a small village called Sankri. A total of 13 mountain peaks of the Central Himalayan Range can be seen from Kedarkantha.

4.7.4   Bali Pass Trek

Lying in the GWS, Uttarkashi, this is the most difficult trek with an altitude of above 5500 m, and it is a nine days trek. The feasible months for trekking are similar to the other treks—May and June months of summer, and September and October months of the post-monsoon season. It is a challenging trek where only trained trekkers can visit. Plenty of exotic flora and fauna are found along the trek. The trek goes across the beautiful Tons River, Siyan Gad (Gad means a small stream), coniferous woodlands, and the magical lake (called 'Ryinsara Tal)'. The other magnificent landscapes are the mountain peaks—Bandarpooch, Kalanag, and Swargarohini.

4.7.5   Auden's Col Trek

A very difficult trek, located at an altitude of about more than 6000 m, the Auden's Col Trek lies in Uttarkashi district. The trek starts from Gangotri and goes to Nala camp through dense pine and cane woodlands. Further, it goes to the Rudra Gaira Base Camp, which connects the Jogin I and Gangotri III peaks and Khatling glacier. The route has spectacular landscapes—glaciers, moraines, narrow cliffs, and snow ridges. For trekking in this route, one has to be physically fit and healthy.

4.7.6   Har Ki Dun Trek

This trek lies in the beautiful Har Ki Dun (Uttarkashi) valley of the Tons River. It stretches both in the valley and the highland of the Tons River's watershed. The maximum altitude of the trek is about 3,800 m. The trek can be covered within seven days. The ideal time to visit this trek is from March to June and September to October. Between November and March, the trail remains covered with huge snow. After the snow-melts in the trail, flowers blossom and the entire region looks beautiful. Plenty of beautiful birds can be seen along the trek. This trek is believed to be sublime.

4.7.7   Satopanth Lake Trek

The trekking can be carried out in June, September, and October and it takes about eight days to complete the trek. The maximum altitude is 4,500 m and the trail is moderately difficult. This trek starts from the Mana village, 3 km from Badrinath Dham (Chamoli district), and ends near the Satopanth Lake. The glacial lake is very beautiful surrounded by the mountain peaks of Satopanth (7,184 m), Chaukhamba (7,138 m), and Balakun (6,471 m). It is a spiritually important trek because it is believed that the trio—Lord Brahma, Lord Vishnu, and Lord Shiva have meditated here, according to our religious wisdom. Further, the entire trek has panoramic landscape features.

4.7.8 The Valley of Flowers Trek

The maximum altitude of the trek is 4,500 m and the trek can be completed within five days. It is an easy trek that can be visited from June to October. Located in the Chamoli district, the Valley of Flowers is declared as a UNESCO Heritage Site. The valley is famous for displaying fragrance wildflowers with rich floral diversity. From the Valley of Flowers, a trek goes to Hemkund Sahib, a famous pilgrimage of Sikhs.

4.7.9 Brahmatal Trek

This trek starts from Lohajung, Deval block in the Chamoli district, opposite to the Roopkund trek. Lohajung is a base camp for Brahmatal, Roopkund, and Bedni Bugyal. The maximum altitude of Brahmatal trek is about 3900 m and about 10–14 days are needed to complete the trek. The ideal time to visit is from December to February therefore, it is called a winter trek. The trek is not much difficult. It has a magnificent view of numerous landscape features. The most important peaks of the central Himalaya—Trishul, Nandaghunti, Chaukhamba, Nilkanth, Kamet, Hathi, and Ghoda can be visualized from this trek. The view of Ali Bugyal, Bedni Bugyal, and Junargali, along with a trail of Roopkund, can also be visualized.

4.7.10 Kalindi Khal Trek

It is the most difficult trek, which covers about a 100 km trail. The trek connects Gangotri and Badrinath pilgrimages. Its course is rough and rugged and it passes through glaciers, snowfields, and rocky moraines. The magnificent peaks of Shivling, Meru, Kedar Dome, Nilkanth, Avalanche Peak, Kamet, and Mana can be visualized along the trek. The maximum height of the trek is 6,500 m altitude. It takes about 15 days to complete the trek and the best time is from June to September.

4.7.11   Roopkund Trek

The Roopkund trek is known as the holy trek because it leads to a lake known as 'Roopkund', which is located on the lap of the Trishul Massif. The lake is mysterious as human skeletons are found in its surroundings. The trek has a spiritual identity because it is related to Lord Shiva and folk deity Goddess Parvati. The maximum height of the trek is 4,800 m and it is a beautiful trek. The famous Ali and Bedni Bugyal lie on the way of the trek. It is one of the best treks in the Garhwal Himalaya. Because of the massive rush on the trail, the High Court of Uttarakhand has banned camping in these meadows.

4.8   Major Trekking Routes of the Kumaon Himalaya

4.8.1   Pindari Glacier Trek

This six-day trek, lying in Bageshwar district (about 110 km), can be visited from March to June and from September to December. The highest altitude of this trek is 4,100 m. Pindari glacier is a source of the Pindar River, which meets with the Alaknanda River at Karnprayag. The trek crosses mountain ridges, spider-walls, and streams with beautiful landscapes.

4.8.2   Kafni Glacier Trek

The Kafni glacier is the source of the Kafni River, a tributary of the Pindar River. The trek of 3 km length bifurcates from the Pindari glacier trek at the Khatiya village. It is an easy trek.

4.8.3   Panchchuli Base Camp

This moderate trek lies in the eastern part of the Kumaon Himalaya, between the Gori Ganga and the Darma Valleys. The 40 km trek starts from Sobla and is rich in alpine meadows, and faunal and floral diversity. Several tribal villages—Dar, Bungling, Sela, Nagling, Baaling, and Dantu are located in Darma valley on the way to Panchchuli Base Camp trek. Along the trek, lush green forests and alpine meadows are found, which make the trek beautiful.

4.8.4 Sunderdhunga Trek

Sunderdhunga trek, located in the Bageshwar district, is a moderate trek in its accessibility. The height of the Sunderdhunga glacier is about 4,300 m. The name Sunderdhunga is comprised of two words—sunder meaning beautiful and dhunga meaning stone thus the trek is called 'the valley of beautiful stones'. It turns to a different direction from the Khati village, continues to go through Jaikholi, Dudhiya Dhaung, and Kathalia villages, and finally reaches the glacier. It is about 24 km from Khati village and acts as a source for the Sunderdhunga River, a tributary of the Pindar River. The four mountain peaks—Tharkot, Maiktoli, Pawali Dwar, and Mrigthuni—can be visualized from the route. The trek is very beautiful above the altitude of 2000 m. April to June and September to October are the suitable months to trek Sunderdhunga and the duration is about 8–10 days.

4.8.5 Nanda Devi East Base Camp

Nanda Devi a beautiful mountain peak and one of the highest peaks of India is located in Chamoli district, Garhwal Himalaya. The Nanda Devi East Base Camp can be reached through two treks. One is from Malari, Chamoli district and the other is from Munsiyari, Pithoragarh district. It is a moderate trek in terms of its accessibility.

4.8.6 Namik Glacier Trek

It is a very tough trek, located in the Pithoragarh district of the Kumaon Himalaya. The Namik Glacier, lying at an altitude of 3,600 m is surrounded by the Nanda Devi, Nanda Kot, and Trishul mountain peaks. In this trek, several waterfalls and natural springs can be seen. A beautiful waterfall 'Birthi' is located en route to the Namik glacier. It is a 40 km trek, which starts from Bala village. Munsiyari, Gogina, and Namik villages are situated on the way to Namik Glacier. The Ramganga River (E) originates from the glacier. The landscape—river valleys, forests, and snow-clad mountain peaks—are panoramic. The area is dominated by Bhotia tribes.

4.8.7 Sinla Pass Trekking

It is also a very difficult trek, located in the Pithoragarh district of the Kumaon Himalaya. On the north of the trek lies Tibet and on the east side, Nepal lies. The trek is very beautiful. The Pithoragarh district is known as the 'Little Kashmir'. The trek is fully covered by snow in the upper watershed and the lower watershed; lush green forests and alpine meadows enhance the beauty of the trek.

4.8.8   Chhota Kailash Trek

The Chhota Kailash trek is situated en-route to the Kailash-Mansarovar in the Vyas valley, Pithoragarh district, Kumaon Himalaya at an altitude of 5,945 m. It bifurcates to Chhota Kailash from Gunji. The trek is mainly devoted to the Hindu pilgrims. The pilgrims who visit Kailash-Mansarovar also visit Chhota Kailash.

The above description on trekking routes of both Garhwal and Kumaon Himalaya shows that the major trekking routes are found in four districts—two districts of the Garhwal Himalaya—Uttarkashi and Chamoli and two districts of the Kumaon Himalaya—Pithoragarh and Bageshwar. A small trekking route is found in the Rudraprayag district of the Garhwal Himalaya. These districts have a major portion covered by snow-clad mountain peaks, which are inaccessible by any means of transportation. However, the snow-clad mountain peaks and the trekking routes have a spectacular landscape. Therefore, these areas have several tourism trekking routes.

4.9   Mountaineering

Mountaineering is the most difficult and life-changing activity practiced in the Uttarakhand Himalaya. The Uttarakhand Himalaya has several world-famous mountain peaks. Among them, Nanda Devi, Trishul, Kamet, and Chaukhamba are famous and suitable for mountaineering. Nanda Devi is the highest mountain peak of Uttarakhand. The mountaineers, in a huge number, practice mountaineering on these mountain peaks.

4.9.1   Mountaineering to Nanda Devi

Nanda Devi (7817 m) is the third-highest mountain peak of India and the fourth-highest mountain peak of the Himalaya. Mountaineers visit Nanda Devi between May and September. Three districts—Chamoli, Pithoragarh, and Bageshwar are located close to the Nanda Devi. NDBR is located just below the Nanda Devi peak. Mountaineering to the Nanda Devi peak can be preceded from Malari in Chamoli district and Munsiyari in Pithoragarh district.

4.9.2 *Mountaineering to Gaumukh*

Gaumukh, a source on the Ganga River, is a major attraction for tourists/pilgrims and mountaineers. The mountaineering route starts from Gangotri town in the Uttarkashi district, a pilgrimage situated 18 km before Gaumukh. The route goes through the bank of the Bhagirathi River with precipitous and undulating terrain. The summer season is ideal for mountaineering to Gaumukh.

4.9.3 *Mountaineering to Panchchuli*

It is a group of mountain peaks that lie in the Pithoragarh district of the Kumaon Himalaya. The altitude of the peak is 6904 m. The months of April–May and September–October are ideal for mountaineering.

4.9.4 *Mountaineering to Om Parvat*

Om Parvat's name is derived from the word 'Om', a pious word for the Hindus. The peak depicting an Om shaped symbol lies on the way to Kailash-Mansarovar in Pithoragarh district.

4.9.5 *Mountaineering to Trishul*

The Trishul peak lies in the Bageshwar district of the Kumaon Himalaya, at an altitude 7122 m. On the foothills of the peak, Roopkund Lake is located. The route has an ideal site for skiing, which is practiced during winter.

4.9.6 *Mountaineering to Chaukhamba*

Chaukhamba peak lies in the Chamoli district of the Garhwal Himalaya. 'Chaukhamba' means four pillars because it has four peaks. The highest peak is located at an altitude of 7138 m. Mountaineering can be done from June to September.

4.9.7   Mountaineering to Bhagirathi Peak

Bhagirathi peak comprises of three peaks—Bhagirathi 1, 2, and 3. The highest altitude is 6856 m, of Bhagirathi Peak 1. This group of peaks lies in the Chamoli district, border with Uttarkashi district. The mountaineering route starts from Mana village and the ideal time to visit is from May to September.

4.10   River Rafting

River rafting is one of the adventure tourism practiced in several rivers of the Uttarakhand Himalaya. The major rivers, where river rafting is practiced, are the Ganga and the Tons in the Garhwal Himalaya and the Kali, Saryu, and the Ramganga (W) in the Kumaon Himalaya.

4.11   River Rafting on the Ganga River

The Ganga River has suitable sites for river rafting and camping between Devprayag and Rishikesh. Some of them are described below.

4.11.1   Rishikesh

Rishikesh is known as the 'Yoga Capital of the World'. The Ganga flows here with gentle slope thus the velocity of Ganga is low. It provides a suitable base for river rafting and camping sites.

4.11.2   Shivpuri

Shivpuri, lying about 14 km of Rishikesh towards Vyasi, is a hub for river rafting. A small service center, it provides suitable camping and river rafting sites.

4.11.3   Byasi

Byasi is a small service center, lying 30 km from Rishikesh towards the Badrinath pilgrimage, on the bank of the holy river Ganga. It is a major camping site for river rafting. Hotels and restaurants are available in this service center, providing ample food and stay for the tourists. The Ganga River flows slowly, which provides a suitable base for river rafting.

4.11.4 Kodiyala

Kodiyala is a small service center located on the left bank of the Ganga River, between Byasi and Devprayag. It is a beautiful service center surrounded by lush green forests. It provides a suitable river rafting and camping sites.

4.11.5 Devprayag

Devprayag is a celestial town which lies at an altitude of about 830 m. The two rivers—Bhagirathi and Alaknanda meet in this town. Devprayag is known as the town of God. The term 'Dev' means God and 'Prayag' means confluence of the two rivers. The Ganga River flows between Devprayag and Rishikesh, providing a suitable base for river rafting.

4.12 River Rafting on the Tons River

4.12.1 Mori

Mori is a small town, called a sleepy town, which lies on the bank of the Tons River in the Uttarkashi district. It has rich cultural diversity and history as it is located in the border area of Uttarakhand and Himachal Pradesh. The world-famous Har Ki Doon valley lies here. River rafting is practiced in the Tons River in Mori town as the major adventure sport in the winter season. Mori town also has suitable camping sites.

4.13 River Rafting in the Kumaon Himalaya

In the Kumaon Himalaya, river rafting is practiced in the Kali, Saryu, and Ramganga (W) Rivers. The Saryu River offers suitable river rafting and camping sites in several places. The river has spiritual importance. River rafting in the Kali River is very adventurous. The Kali River has high volume and velocity of water however wherever, the velocity of the river is slow, river rafting is practiced. Pithoragarh is famous for river rafting in the Kali River. River rafting is practiced in several places along the Ramganga River. The river provides ample and suitable sites for river rafting and camping. Boating is practiced in the Naini Lake, Bhimtal, and Naukuchiyatal in the Nainital district.

4.14 Skiing

Auli is a world-famous skiing destination. Skiing is played as a winter sport where the players of skiing come from all around the world. Auli is alpine pastureland that lies about 15 km from Joshimath in Chamoli district of the Garhwal Himalaya.

Besides all these types of tourism, rural/village tourism is emerging in the Uttarakhand Himalaya. Here, many villages are located close to the snow-clad mountain peaks, which have a high potential for tourism development. One of the examples is Munsiyari village in Pithoragarh district. The villages have a rich traditional culture, nutritional food, and beverages, which the tourists enjoy during their stay.

4.15 Purpose of Tourism and Tourists/Pilgrims' Inflow

As discussed in the preceding paragraphs, the Uttarakhand Himalaya practices various types of tourism where exodus number of tourists/pilgrims visits every year. Table 4.1 shows the visitors and their purpose of visiting the Uttarakhand Himalaya.

Table 4.1 Visitors and their purpose of visiting the Uttarakhand Himalaya

Purpose of visit	Domestic (%)	Foreign (%)	Total visitors (%)
Cultural activities	53.1	59.3	53.8
Resorts	13.9	10.0	13.5
Wildlife	10.7	10.2	10.6
Others	8.3	4.7	7.9
Visiting friends and family	2.9	2.3	2.8
Business and professional	4.6	1.2	4.2
Health and treatment	0.6	0.6	0.6
Social/religious function	5.5	11.6	6.2
Total	100.0	100.0	100.0

Source UTDB, Dehradun, 2019

The highest number of visitors both domestic and foreign that come here for cultural activities, including pilgrimages, is 53.8%, followed by the tourists that come for summer resorts (13.5%), and wildlife (ecotourism) (10.6%). Others are here to visiting friends and families, or for business and professional purposes, or health and treatment, and or social/religious functions, which are fewer number.

4.16 Conclusions

In this chapter, the major types of tourism and tourist/pilgrim centers have been described. The natural locales and pilgrimage centers have been briefly described whereas, the major trekking and mountaineering routes have been described broadly. A table has been presented on visitors and their purpose of visiting the Uttarakhand Himalaya. It reveals that a large proportion of visitors, more than 50% both domestic and foreign, visit for cultural activities including pilgrimages. The description of tourism types and tourist centers further shows that Uttarakhand Himalaya has tremendous potential for sustainable tourism development and if the proper infrastructural facilities are developed and provided to visitors, it can gain an impressive position in tourism development.

References

Government of India (2008) Uttarakhand Tourism Development Master Plan 2007–2022. https://uttarakhandtourism.gov.in/utdb/sites/default/les/volume-3-appendices.pdf. Accessed on 1 Dec 2016

Kohli MS (2002) Mountains of India: tourism, adventure & pilgrimage. Indus Publishing Company, New Delhi

Nakajo A (2017) Development of tourism and the tourist industry in India: a case study of Uttarakhand. J Urban Region Stud Contemp India 3(2):1–12

Nakatani T (2011) Indo: Seichijunrei, Sei-Zoku-Yuu. (Tourism in India: Pilgrimage, Sacredness, Secularity, Amusement). In: Yasumura K et al (ed) Yoku Wakaru Kanko-Shakaigaku (Introduction of Tourism Sociology) Minerva Shobo, pp 180–181

Sati VP (2015) Pilgrimage tourism in mountain regions: socio-economic implications in the Garhwal Himalaya. South Asian J Tour Heritage 8(1):164–182

Sati VP (2013) Tourism practices and approaches for its development in the Uttarakhand Himalaya India. J Tour Challenges Trends 6(1):97–112

Sharma JK (2008) Types of tourism and ways of recreation. Kanishika Publishers, New Delhi

Siddiqui S (2000) Eco-friendly tourism in U. Publishing, Delhi, P. Himalaya. B. R

Chapter 5
The Trends of Tourism and Tourists/Pilgrims' Inflow

5.1 Introduction

Tourism is an activity in which people travel for leisure, business, education, and performing rituals. It is practiced for leisure, recreation, entertainment, education, and culture (Tribe 2009; Smith and Jenner 1997; Kelman and Doods 2009; Ellis 2003; Kulendran and Witt 2003). Tourism has different types such as nature tourism, pilgrimage tourism, and adventurous tourism. It is fast-growing as smokeless industry and a major source of income and economy for the people worldwide (Sharpley 2004). India is bestowed with tremendous potential of tourist/pilgrim destinations in the forms of natural locales, pilgrimages, historical monuments, rich culture, and natural parks and wildlife sanctuaries. Every year, an exodus number of tourists visit these places of tourists/pilgrims' interests. India is home to 36 world heritage sites, which were visited by over 8.80 million foreign tourists in 2016 (Government of India 2013). WTTC (2017) reported that India generated USD 220 billion from the tourism industry, which is 9.6% of the national GDP. Globally, tourism is a USD 625 billion industry, the single largest non-government economic sector in the world (WTO 2005).

Uttarakhand, nestled on the lap of the Himalaya has spectacular landscapes—river valleys, middle altitudes, highlands, alpine pastures, and the snow-clad mountain peaks. It has world-famous highland and the river valley pilgrimages—Badrinath, Kedarnath, Yamunotri, Gangotri, Rishikesh, and Haridwar; natural locales—Dehradun, Mussoorie, Nainital, Ranikhet, Almora, and Kausani; national parks and wildlife sanctuaries, and many other places of tourists' and pilgrims' interest (Sati 2013). Because of the diverse nature of tourist places, pilgrimage tourism, natural tourism, and eco-tourism are practiced in the Uttarakhand Himalaya (Sati 2015). Besides, adventure tourism in the form of mountaineering, trekking, skiing, and river rafting are practiced here. The pilgrims, within and outside India, believe Uttarakhand to be the 'land of the Gods and Goddesses'. The Ganga, which originates and flows

V. P. Sati, *Sustainable Tourism Development in the Himalaya: Constraints and Prospects*, Environmental Science and Engineering, https://doi.org/10.1007/978-3-030-58854-0_5

from Uttarakhand, is called the mother Ganga, which is one of the major attractions for the pilgrims (Sati 2019).

Pilgrimage to the highland and river valley sacral places in Uttarakhand is a centuries-old practice and is still very popular. Out of the total tourists/pilgrims that arrive in Uttarakhand, 80% of them visited these pilgrimages (Sati 2018). Haridwar and Rishikesh pilgrimages received more than 50% of the total tourists/pilgrims arriving in Uttarakhand. The highland pilgrimages are located in spectacular landscapes, mainly in the alpine pasturelands (locally known as Bugyals), where the terrain is very fragile and highly prone to natural disasters. Badrinath and Gangotri pilgrimages are connected by all-weather roads and for Kedarnath and Yamunotri, the pilgrims have to trek about 16 km. Meanwhile, Rishikesh and Haridwar are located in the plain areas of the Ganga valley, well connected by air, rail and roadways.

Tourism is a large industry in Uttarakhand as it generates 50% revenues of the SDGP. It has tremendous potential for tourism development as tourists/pilgrims' inflow is significantly high. During the last 18 years (2000–2018), the total number of tourists/pilgrims visited Uttarakhand was 434.03 million, of which, 300 million domestic pilgrims and 0.48 million foreign pilgrims visited the major pilgrimages. Haridwar was visited by about 240 million domestic and 0.32 million foreign pilgrims. About 56.1 million domestic tourists and 0.5 million foreigners visited the natural locales, and 76 million domestic and 0.95 million foreign tourists visited the administrative tourist places.

Although the Uttarakhand Himalaya has plenty of world-famous tourists' places and pilgrimages yet, it does not have an impressive position in tourism development. The entire region is lagging in infrastructural facilities such as transportation, accommodation, and institutions. The UTDB is a state government agency, whose role is to provide basic amenities to tourists/pilgrims however, it has not done any remarkable work in this regard. Also, the state of Uttarakhand struggles to harness the huge potential of tourism. There have been lots of studies conducted by the scholars on tourism aspects, however, a detailed study of tourists/pilgrims' inflow for the two decades, and specific tourism types and their precise study had not been carried out so far. This chapter examines the nature and trends of tourism and tourists/pilgrims inflow in the Uttarakhand Himalaya from 2000 to 2018. It studies the types of tourism and inflow of tourists/pilgrims in the tourist places—natural locales, pilgrimages, and administrative towns/cities. It also analyses the tourists/pilgrims' inflow both domestic and international and illustrates the changing trends of tourists/pilgrims' inflow from year to year between 2000 and 2018. The Uttarakhand Himalaya has numerous tourist places and pilgrimages however for this study, only 27 tourist places and pilgrimages have been studied.

Data on tourists/pilgrims' inflow, both domestic and foreign, to 27 natural locales, administrative towns/cities, and pilgrimages were collected from 2000 to 2018. Descriptive statistics were used to analyze the minimum, maximum, mean value, and standard deviation of tourists/pilgrims' inflow in the major places of tourists' interest. The trend of tourists/pilgrims' inflow has been presented using graphs. First, the tourists/pilgrims' inflow (domestic and foreign separately) in three different tourist places have been presented in graphs. Thereafter, the data were grouped into two types—tourists/pilgrims' inflow (domestic and foreign separately) in the highland and river valleys pilgrimages (total six) and combined natural locales and administrative towns/cities (total five). The data has been largely illustrated.

5.2 Trends of Tourism and Tourists/Pilgrims' Inflow

In this section, tourist places are classified as natural locales, cultural places (pilgrimages), and administrative towns/cities. Further, tourists/pilgrims' inflow, both domestic and foreign, is shown from 2000 to 2018. A detailed description of tourists/pilgrims' inflow in these tourists/pilgrims centers are as follows.

5.3 Domestic Tourists/Pilgrims' Inflow

Domestic tourists/pilgrims' inflow in the natural locales, cultural places, and administrative towns/cities are shown in Fig. 5.1. Pilgrims' inflow is the highest in cultural places and it has increasing trends. Tourist/pilgrims' inflow data of eight major pilgrimage centers, 10 major administrative towns/cities, and nine major natural locales were gathered from UTDB, Dehradun from 2000 to 2018. In 2000, domestic pilgrims' inflow in the major pilgrimages was 7.41 million, which increased to 25.26 million in 2018. In 2013, the entire Uttarakhand was affected by cloudburst triggered natural calamity therefore, the pilgrims' inflow in the pilgrimages decreased to 14.88 million whereas, in 2012, it was 19.32 million. Meanwhile, the growth in the number of pilgrims after 2013 increased continuously. The total domestic pilgrims, who visited the pilgrimages, were 300 million. Domestic tourists' inflow in administrative towns/cities increased from 1.97 million in 2000 to 6.7 million in 2018, which was less than the pilgrims' inflow. Similarly, tourists' inflow in administrative towns/cities decreased from 5.07 million to 3.77 million in 2013, and then continuously increased in the later years. The total domestic tourists' inflow in administrative towns/cities was 76 million. Domestic tourists' inflow in the natural places

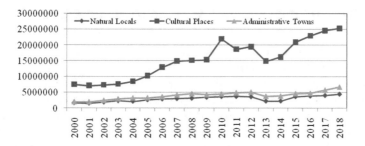

Fig. 5.1 Domestic tourists' inflow

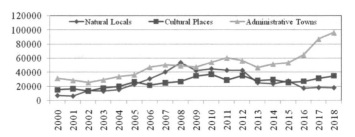

Fig. 5.2 Foreign tourists' inflow

was the lowest among all three places. In 2000, tourist's inflow was 1.64 million, which increased to 4.48 million in 2018. However, in 2013, it decreased to 2.21 million while it was 3.62 million in 2012. After 2013, the growth in tourists' inflow in natural places doubled. A total of 56.09 million domestic tourists visited in the natural locales during the mentioned 18 years.

5.4   Foreign Tourists/Pilgrims' Inflow

Foreign tourists/pilgrims' inflow was highest in the administrative towns/cities, which was 0.95 million (Fig. 5.2). It was followed by natural locales (0.5 million) and further followed by pilgrimages (0.48 million). In 2000, foreign tourists who visited the administrative towns/cities were 0.32 million, which increased to 0.96 million in 2018. In between, there was no clear cut trend. Foreign tourists' inflow decreased in 2001, 2002, 2008, 2009, and 2013. Further, its inflow for pilgrimage tourism was 0.15 million, which reached 0.35 million in 2018 (just double) with a substantial decrease in 2002, 2006, 2011, 2013, and 2015. In the natural locales, the foreign tourists' inflow was 0.071 million in 2000. It increased to 0.18 million in 2018. The number decreased in 2009, 2013, and 2014. Therefore there was no straight trend of foreign tourists' inflow.

5.5   Domestic Pilgrims' Inflow in the Highland Pilgrimages

The pilgrimages of Uttarakhand are divided into two categories—the highland pilgrimages, and the river valley pilgrimages. Data on individual pilgrimage centers were gathered and have been described. Yamunotri, Gangotri, Kedarnath, and Badrinath (located west to east) are the highland pilgrimages, where a large number of pilgrims visit every year. During 2000–2018, the highest number of domestic pilgrims visited Badrinath (13.1 million), followed by Kedarnath (7.03 million), Gangotri (5.07 million), and Yamunotri (4.23 million). In Badrinath, the highest number of pilgrims visited in the years 2008 (1.1 million), 2012 (1.04 million), and 2018 (1.04

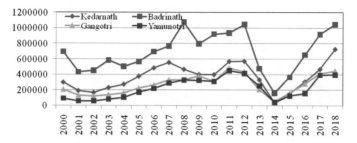

Fig. 5.3 Domestic pilgrims' inflow in the highland pilgrimages

million). The domestic pilgrims' inflow decreased in 2001, 2009, 2013, and 2014. In Kedarnath, the highest number of pilgrims visited in the years—2018 (0.73 million), 2012 (0.57 million), 2011 (0.56 million), and 2007 (0.55 million). In 2001, 2002, 2013, and 2014, pilgrims' inflow decreased. In Gangotri, the lowest pilgrims' inflow was 0.051 million in 2014 and the highest inflow was 0.48 million in 2011. Similarly, in Yamunotri, the lowest pilgrims' inflow was 0.038 million in 2014 and the highest inflow was 0.44 million in 2011 (Fig. 5.3).

5.6   Foreign Pilgrims' Inflow in the Highland Pilgrimages

Foreign pilgrims' inflow in the highland pilgrimages is comparatively low. The Kedarnath pilgrimage received the highest inflow, a total of 23,094 from 2000 to 2018, followed by Gangotri (6429), Badrinath (3805), and Yamunotri (3752). In 2000, foreign pilgrims' inflow was zero in all the highland pilgrimages. In Kedarnath, the highest pilgrims' inflow was 4811 in the year 2005 and the lowest inflow was 81 in the year 2013. The Badrinath pilgrimage received the highest pilgrims inflow (1064) in the year 2018 and the lowest was 12 in 2006 (Fig. 5.4). The highest pilgrims' inflow in the Gangotri pilgrimage was 1109 in 2016 and the lowest was in the year 2014, which was 139. The Yamunotri pilgrimage received the lowest foreign pilgrims' inflow in 2002 (47) and the highest inflow in 2018 (482).

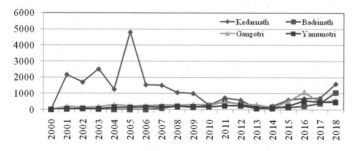

Fig. 5.4 Foreign pilgrims' inflow in highland pilgrimages

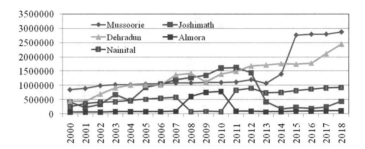

Fig. 5.5 Domestic tourists' inflow in the major natural/administrative centers

5.7   Domestic Tourists' Inflow in the Major Natural Locales/Administrative Towns

Five out of a total of 19 tourist places of natural locales and administrative towns were selected for the detailed description because of them being the major centers with high tourists' inflow. Figure 5.5 shows five major centers of tourists' interests—Mussoorie, Dehradun, Nainital, Joshimath, and Almora. The total tourists' inflow in Mussoorie was 27.22 million, followed by Dehradun (25.64 million), Nainital (10.54 million), Joshimath (14.25 million), and Almora (3.53 million). The trend of growth in terms of the number of tourists in these centers was stagnant up to 2007 with the highest inflow in Mussoorie, followed by Dehradun, Joshimath, Nainital, and Almora. Both Mussoorie and Dehradun received an increase in tourists' inflow after 2014, which continued increasing in the later years.

5.8   Foreign Tourists' Inflow in the Major Natural Locales/Administrative Towns

Foreign tourists' inflow in the major natural locales/administrative towns in the Uttarakhand Himalaya were about 10 times less than the domestic tourists. Further, the trend of tourists' inflow remained almost unchanged from 2000 to 2018 except for the tourists' inflow in Dehradun city, which kept increasing and was the highest. There was an increase in tourists' inflow in Joshimath town between 2005 and 2009, and after 2013, the tourists' inflow decreased continuously. In Nainital, tourists' inflow decreased in 2008, 2009, and 2010. During the period, Dehradun received the highest tourists' inflow (0.34 million), followed by Nainital (0.12 million), and Almora (0.083 million). Mussoorie has received 0.082 million and Joshimath has received 0.061 million tourists (Fig. 5.6).

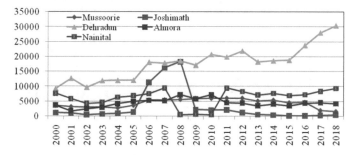

Fig. 5.6 Foreign tourists' inflow in the major natural/administrative tourist destinations

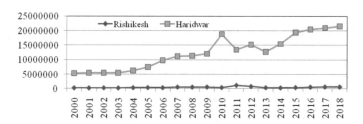

Fig. 5.7 Domestic tourists' inflow in the river valley pilgrimages

5.9 Domestic Pilgrims' Inflow in the River Valleys Pilgrimages

The two river valley pilgrimages—Rishikesh and Haridwar are described. These two pilgrimages are world-famous, where thousands of pilgrims visit every year. Figure 5.7 shows trends of domestic pilgrims' inflow in these pilgrimages. In Haridwar, the trend kept increasing, except in the year 2013, when pilgrims' inflow decreased. The Rishikesh pilgrimage had unchanged tourists' inflow during the period. The total domestic pilgrims' inflow in Rishikesh was 91.2 million, and in Haridwar, it was 240 million, more than half of the total pilgrims' inflow in the Uttarakhand Himalaya.

5.10 Foreign Pilgrims' Inflow in the River Valleys Pilgrimages

Haridwar has received a total of 0.32 million foreign pilgrims, three and a half times more than that of Rishikesh (0.097 million). The trend of foreign pilgrims' inflow in Haridwar kept on increasing until 2010. It decreased after 2010, followed by an irregular trend. In terms of Rishikesh, the trend remained unchanged and straight (Fig. 5.8).

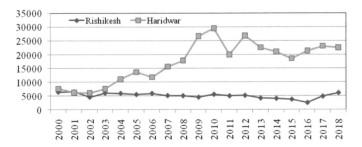

Fig. 5.8 Foreign tourists' inflow in the river valleys pilgrimages

5.11 Analysis of Tourists/Pilgrims' Inflow Using Descriptive Statistics

Table 5.1 shows the descriptive statistics of tourists/pilgrims' inflow in different tourist/pilgrim places of the Uttarakhand Himalaya. The lowest tourists'/pilgrims' inflow was in natural locales (1.5 million), while the maximum inflow of tourists/pilgrims was in cultural places (25.3 million). It shows that the mean of tourists/pilgrims' inflow was the highest in cultural places, followed by the inflow in administrative towns, and the lowest inflow was in natural locales. The minimum, maximum, and mean value of foreign tourists/pilgrims' inflow is given. The maximum inflow was noticed in the administrative towns, comprising of 96,026 tourists, followed by the inflow in natural places, which was 53,419, and in the pilgrimages, the inflow was the lowest (36,777). The lowest minimum inflow was in natural locales and the highest minimum inflow was in the administrative towns i.e., 25,448 in number. Therefore, comparatively low tourists/pilgrims' inflow was observed being highest in the administrative towns, which are located in the plains and easily accessible areas. The minimum value of domestic pilgrims' inflow was noticed in Yamunotri (38,208) followed by Kedarnath (40,718). The maximum inflow of pilgrims was in Haridwar, which was more than 21.56 million. Likewise, the lowest mean value of tourists' inflow was (266,885) in Gangotri and the highest was in Haridwar (12.54 million). The mean value of foreign pilgrims inflow in pilgrimages was the highest in Haridwar with 17,360 pilgrims while the lowest mean value of pilgrims' inflow was in Yamunotri (197 pilgrims). Likewise, the highest maximum inflow was in Haridwar (29,555 pilgrims) and the lowest maximum inflows were in Yamunotri (566 pilgrims). Minimum foreign pilgrims' inflow was zero in all the four highland pilgrimages. The analysis shows that foreign pilgrims mainly visited the river valleys pilgrimages. The mean value of domestic tourists' inflow was the highest in Mussoorie (1.43 million tourists), followed by Joshimath (0.75 million tourists). The lowest mean value of domestic tourists' inflow was in the valley of flowers (5642 tourists), followed by Auli (33,067 tourists). Similarly, the minimum and maximum tourists' inflow was changed. Chilla is a tourist place, located in the RJNP, where the highest number of foreign tourists' visited (26,914), followed by Joshimath (26,914). The lowest foreign tourists' inflow was in Auli (561), followed

Table 5.1 Domestic/foreign tourists/pilgrims' inflow

Domestic tourists/pilgrims' inflow in different tourists/pilgrims places ($n = 19$)

Tourism type	Minimum	Maximum	Mean	Std. deviation
Natural locales	1,500,187	4,478,338	2,952,310	881,995
Cultural places	7,048,942	25,262,167	15,297,309	6,144,652
Administrative towns	1,942,542	6,701,955	4,004,556	1,226,359

Foreign tourists/pilgrims' inflow in different tourists/pilgrims places ($n = 19$)

Tourism type	Minimum	Maximum	Mean	Std. deviation
Natural locales	6324	53,419	26,488	13,934
Cultural places	12,955	36,777	25,609	7136
Administrative towns	25,448	96,026	49,905	18,577

Domestic pilgrims' inflow in pilgrimages $N = 19$

Pilgrimages	Minimum	Maximum	Mean	Std. deviation
Kedarnath	40,718	730,387	370,423	174,270
Badrinath	159,405	1,075,372	688,669	259,243
Gangotri	51,555	484,826	266,885	125,845
Yamunotri	38,208	448,751	222,387	137,599
Rishikesh	220,097	1,181,535	480,038	240,164
Haridwar	5,316,980	21,555,000	12,536,162	5,742,652

Foreign pilgrims' inflow in pilgrimages $N = 19$

Pilgrimages	Minimum	Maximum	Mean	Std. Deviation
Kedarnath	0	4811	1215	1112
Badrinath	0	1064	200	252

(continued)

Table 5.1 (continued)

Domestic tourists/pilgrims' inflow in different tourists/pilgrims places ($n = 19$)				
Gangotri	0	1109	338	235
Yamunotri	0	566	197	155
Rishikesh	2574	6536	5092	1005
Haridwar	6029	29,555	17,360	7392

Domestic tourists' inflow in natural locales/administrative towns $N = 19$				
Natural locales	Minimum	Maximum	Mean	Std. deviation
Mussoorie	847,191	2,870,475	1,432,772	736,516
Srinagar	38,391	324,218	175,562	74,882
Chilla	99,102	439,034	282,302	85,449
Joshimath	173,013	1,626,275	749,901	528,555
Auli	6459	151,560	33,067	34,398
Valley of Flower	176	14,128	5642	4542
Ranikhet	62,487	150,423	90,830	31,601
Kausani	62,485	191,866	90,298	43,304
Kathgodam	40,642	162,087	91,934	47,614

Foreign tourists' inflow in natural locales/administrative towns $N = 19$				
Natural locales	Minimum	Maximum	Mean	Std. deviation
Mussoorie	1550	5985	4298	1471
Srinagar	69	5192	1220	1664
Chilla	13	26,914	14,146	7886
Joshimath	155	18,252	3233	5516

(continued)

Table 5.1 (continued)

Domestic tourists/pilgrims' inflow in different tourists/pilgrims places ($n = 19$)

Auli	93	561	295	131
Valley of Flower	5	763	336	221
Ranikhet	398	1683	749	375
Kausani	303	9001	1834	2955
Kathgodam	19	1360	378	284

Source By author

by the Valley of Flowers (763). In terms of the minimum foreign tourists' inflow, the lowest number of tourists visited the Valley of Flowers (5), followed by Auli (93). The highest minimum foreign tourists visited in Mussoorie (1550), followed by Ranikhet (398), and Kausani (303). The mean value of tourists' inflow changed accordingly.

5.12   Discussion

It has been noticed that the pilgrims which visited the major pilgrimages are outnumbered mainly the domestic pilgrims. The reason is that the Uttarakhand Himalaya is known as the 'Land of Gods and Goddesses'. It has many highlands and river valley pilgrimages where exodus number of pilgrims visit every year mainly the Hindu pilgrims. The pilgrims believe that visiting these highland pilgrimages once in a lifetime will help them get rid of the cycle of birth and death. Although there are several natural locales and administrative towns/cities of tourist interests, and tourists visit these tourist places throughout the year yet, the inflow of tourists is less than the inflow of pilgrims who visit the pilgrimages, both the river valleys and the highlands.

On the other hand, the number of foreign pilgrims that visit pilgrimages is quite less. Therefore, it has been observed that foreign tourists' inflow is higher in the administrative cities/towns and natural locales of tourists' interests than in the major pilgrimages. The foreign pilgrims visiting the pilgrimages are mainly from south and southeastern Asia and few are from the other parts of the world. However, foreign tourists visiting tourist places come from nations almost from all corners of the world.

The largest number of domestic pilgrims visited the Badrinath Pilgrimage. Badrinath pilgrimage is connected by road (all-weather) therefore, the number of pilgrims visiting here is high. Kedarnath pilgrimage ranks second in pilgrims' inflow because it is well connected by airways, where chopper facility is provided to pilgrims from Jolly Grant, Dehradun, and Phata (Guptakashi). The pilgrims who do not travel by air, trek 16 km to reach Kedarnath. Badrinath is famous for the Vishnu temple while Kedarnath is famous for the Shiva temple, and pilgrims of both sects—Shaiv and Vaishnav visit these temples. Gangotri and Yamunotri are the other two highland pilgrimages where Shakti is worshiped. Gangotri is well connected by roads whereas, Yamunotri can be reached by trekking of about 16 km. It has been noticed that in the pilgrimages, which are well connected, pilgrims' inflow is high. In terms of foreign pilgrims' inflow in the highland pilgrimages, it is the highest in Kedarnath because of its connectivity by air. Foreign pilgrims' inflow in other pilgrimages is low.

It has been noticed that pilgrims' inflow in these highland pilgrimages decreased mainly during the catastrophic natural disasters. In 2013, the highland pilgrimages were devastated due to cloudbursts-triggered debris flows and flash floods. The entire Uttarakhand Himalaya is highly vulnerable to catastrophic natural disasters (Sati 2014) and as a result, the pilgrims' inflow decreases when natural disasters occur.

The Uttarakhand Himalaya is endowed with numerous natural locales of tourist interests. Among the major natural locales and administrative towns the author studied, the highest tourists' inflow was noticed in Mussoorie and Dehradun and the trend has been increasing. Dehradun is the capital city of Uttarakhand, which lies in the serene valley of *Doon*. The Ganga River flows in its east part and the Yamuna River flows in the west, enhancing its beauty. Further, the city is well connected by air, rail, and roads. These drivers promote tourism in Dehradun. Mussoorie is another important destination where several tourists visit every year. Tourists' inflow in Almora is comparatively less because of its location and inaccessibility although, it is a famous tourist destination. Nainital and Joshimath are the other famous tourist destinations. Joshimath is located on the way to Badrinath and it is the winter home of Lord Vishnu therefore, tourists' inflow is high. Dehradun is also a hotspot for foreign tourists. The trend of tourists' inflow has been increasing. The foreign tourist inflow in other tourists' destination remained the same during the period with the highest inflow in Nainital, followed by Mussoorie and Almora. Joshimath has the lowest inflow of foreign tourists whereas from 2005 to 2009, the inflow in Joshimath was the highest.

Haridwar and Rishikesh are two world-famous river valley pilgrimages, where hundreds of thousands of pilgrims visit from India and the world. Both pilgrimages are accessible by road, rail, and air transports. Haridwar is known as the gateway to Badrinath and Kedarnath. Hari Ki Pauri is the place where the Ganga Aarti is performed. The Ganga River enters in the plain region from Haridwar. The pilgrims believe that taking a bath in the Hari Ki Pauri once in a lifetime help them get rid of the cycle of birth and death, and their forefathers rest in heaven after death. Rishikesh is the world's 'Yoga Capital' and a place for spiritual tourism. Both Haridwar and Rishikesh has almost become twin cities expect a small strip of RJNP in Raiwala. Two-third of the total pilgrims visits Rishikesh and Haridwar every year.

The tourism trend is not uniformed. The tourists/pilgrims' inflow was quite less mainly in the years when the Uttarakhand Himalaya was affected by natural disasters. This part of the Himalaya receives heavy downpour called cloudbursts. Cloudburst-triggered debris flows and flash floods are common and devastating. The pilgrimages/tourist places are located in the fragile/vulnerable landscapes and river valleys and thus, during the devastating atmospheric events, the tourists/pilgrims inflow decreases.

5.13 Conclusions

The nature of tourism and tourists/pilgrims' inflow has been described in this chapter. The study revealed that the pilgrims' inflow in the pilgrimages was higher than tourists' inflow in the natural locales. Further, the pilgrims visiting the river valleys pilgrimages were higher in number than those visiting the highland pilgrimages. The trend of tourists/pilgrims' inflow was not uniform in all the pilgrimages and natural locales. The inflow decreased during the occurrence of natural disasters, mainly in the highlands pilgrimages and natural locales. However, tourists/pilgrims' inflow has increased from year to year. Infrastructural facilities—transportation, accommodation, and institutions can be developed in all the pilgrimages and natural locales thus, tourists/pilgrims' inflow may be increased, which can enhance the income, economy, and livelihood of the people and the state.

References

Ellis C (2003) When volunteers pay to take a trip with scientists—participatory environmental research tourism (PERT). Human Dimen. Wildlife 8(1):75–80

Government of India (2013) India tourism statistics at a glance, Ministry of Tourism, Government of India. https://tourism.gov.in/sites/default/files/Other/Incredible%20India%20final%2021-7-2014%20english.pdf. Accessed on 1 May 2017

Kelman I, Dodds R (2009) Developing a code of ethics for disaster tourism. Int J Mass Emerg Disasters 27(3):272–296

Kulendran N, Witt SF (2003) Forecasting the demand for international business tourism. J Travel Res 41(3):265–271

Sati VP (2019) Himalaya on the threshold of change. Springer International Publishers, Switzerland, p 250. 10.1007/978-3-030-14180-6

Sati VP (2018) Carrying capacity analysis and destination development: a case study of gangotri tourists/pilgrims' circuit in the Himalaya. Asia Pacific J Tour Res APJTR 23(3):312–322. https://doi.org/10.1080/10941665.2018.1433220

Sati VP (2015) Pilgrimage tourism in mountain regions: socio-economic implications in the Garhwal Himalaya. South Asian J Tour Heritage 8(1):164–182

Sati VP (2014) Landscape vulnerability and rehabilitation issues: a study of hydropower projects in the Garhwal region, Himalaya. Nat Hazards 75(3):2265–2278. https://doi.org/10.1007/s11069-014-1430-y

Sati VP (2013) Tourism practices and approaches for its development in the Uttarakhand Himalaya India. J Tour Challenges Trends 6(1):97–112

Sharpley R (2004) Tourism: a vehicle for development? In: Sharpley R, Telfer DJ (eds) Tourism and development: concepts and issues. Channel View Publications

Smith C, Jenner P (1997) Educ Tour Travel Tour Anal 3:60–75

Tribe J (2009) Philosophical issues in tourism. Channel View Publications, Bristol, United Kingdom

World Tourism Organization (2005) Making tourism more sustainable: a guide for policy makers, WTO Publications

WTTC (2015) Travel and tourism economic impact 2015 Indonesia, The authority on World Travel and Tourism, World Travel and Tourism Council, on-line edition; rochelle.turner@wttc.org

Chapter 6
Major Tourists/Pilgrims' Circuits

6.1 Introduction

The Uttarakhand Himalaya has numerous places of tourists/pilgrims' interests. The places are categorized as natural locales, pilgrimages, administrative towns/cities, and national parks and wildlife sanctuaries (Sati 2013). The author has divided the Uttarakhand Himalaya into tourists/pilgrims' circuits very comprehensively based on the nature of tourist places and their locations. Further, the geographical and cultural components of tourism were taken into consideration at the time of delineating them. Each circuit has its characteristics. The circuits are either situated along the river valleys, or in the Middle Himalaya and or alpine pasturelands. Some circuits are extended in two districts of two river valleys. Suppose the Panch Kedar circuit lies in both Rudraprayag and Chamoli districts and both Mandakini River valley and Alaknanda River valley (Sati 2015). Dehradun, Mussoorie, Chakrata, Rishikesh, and Haridwar circuit is also another example of it. This circuit falls in Dehradun and Haridwar districts. Further, it comprises the Ganga valley and the Yamuna valley. The circuits may be cultural, natural or adventure, or a combination of all these three types. The Uttarkashi and Gangotri tourist circuit is the best example of the natural, cultural, and adventurous type (Sati 2018). Chilla-Pauri tourist circuit is an example of a natural, park, and wildlife type. Similarly, Mukteshwar-Rudrapur is characterized as a natural, cultural, and park and wildlife tourist circuit. These tourists/pilgrims' circuits cover all the tourist places of the Uttarakhand Himalaya. Although the UTDB has divided the Uttarakhand Himalaya into tourism circuits yet these circuits do not cover all the places of tourists' interest. The main purpose of dividing the Uttarakhand Himalaya into tourists/pilgrims' circuit was to make a suitable policy at the level of tourist circuit for sustainable tourism development in the state. The description of each tourist/pilgrim circuit is empirical, mainly based on the observation/experience of the author after he visited the entire Uttarakhand Himalaya. Data on tourists/pilgrims' inflow on each circuit were gathered from UTDB and

V. P. Sati, *Sustainable Tourism Development in the Himalaya: Constraints and Prospects*, Environmental Science and Engineering, https://doi.org/10.1007/978-3-030-58854-0_6

analyzed by the author. This chapter inclusively describes these circuits. Photographs
of the tourist places within the circuits and a map showing the major tourist circuits
of the Uttarakhand Himalaya further enrich the contents.

6.2 Major Tourists/Pilgrims' Circuits

The author has divided the Uttarakhand Himalaya into nine tourist/pilgrim circuits,
out of which, six circuits are situated in the Garhwal Himalaya and three circuits
are situated in the Kumaon Himalaya. The Garhwal Himalaya is well-known
for its world-famous pilgrimages—the highlands and the river valleys; besides,
natural locales and wildlife sanctuaries are also located here. In the meantime,
the Kumaon Himalaya has several natural locales and CNP. Many natural locales,
pilgrimage centers, parks and sanctuaries, and trekking routes are located close
to each other. Therefore, many tourists/pilgrims' circuits are a mixture of natural,
cultural, adventure, and wildlife (ecotourism) categories (Fig. 6.1; Table 6.1).

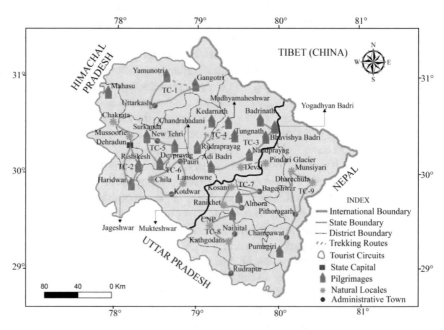

Fig. 6.1 Major tourists/pilgrims' circuits in the Uttarakhand Himalaya

Table 6.1   Major tourists/pilgrims' circuits

S. No.	Tourists/pilgrims' circuits	Nature of circuit	Districts
TC 1	Uttarkashi-Yamunotri-Harsil-Gangotri	Natural, cultural, and adventure	Uttarkashi
TC 2	Dehradun-Mussoorie-Chakrata-Rishikesh-Haridwar	Natural and cultural	Dehradun and Haridwar
TC 3	Panch Badri: Badrinath, Yogadhyan Badri, Bridha Badri, Bhavishya Badri, and Adi Badri; Vishnuprayag, Nandprayag and Karnprayag	Cultural	Chamoli
TC 4	Panch Kedar: Kedarnath-Tungnath-Rudranath-Kalpeshwar-Madhyamaheshwar; Rudraprayag and Sonprayag	Cultural and adventure	Rudraprayag and Chamoli
TC 5	Devprayag-Kunjapuri-Chandrabadni-Surkanda-New Tehri	Cultural and natural	Tehri
TC 6	Chilla-Kotdwar-Lansdowne-Pauri	Natural and park and wildlife	Pauri
TC 7	Ranikhet-Almora-Jageshwar-Kausani-Baijnath-Bageshwar	Natural and cultural	Almora and Bageshwar
TC 8	Mukteshwar-Nainital-Bhimtal-Corbett-Rudrapur	Natural, cultural and park and wildlife	Nainital and USN
TC 9	Champawat-Pithoragarh-Dharchula-Munsiyari	Natural and cultural	Champawat and Pithoragarh

6.3   Uttarkashi-Yamunotri-Harsil-Gangotri Circuit

This circuit lies in the Uttarkashi district. It has many places of tourists/pilgrims interest—the highland pilgrimages and natural locales. Uttarkashi—a natural and cultural town lying on the bank of the holy Bhagirathi River, is the central place of this circuit. Uttarkashi is believed to be an old town where Lord Vishwanath temple lies. Both tourists and pilgrims visit here to proceed to Yamunotri and Gangotri pilgrimages (Fig. 6.2). From Uttarkashi, the first route goes to Yamunotri pilgrimage through Dharasu. Road transport is available from Dharasu to Kharsali, which is about 60 km. Kharsali is the winter resort of the Goddess Yamunotri. Yamunotri is a 16 km trek from Kharsali. On the way, a place named Garam Sota (hot spring) is located. The pious Yamuna River originates from Yamunotri, which flows towards the southwest and meets the Tons River in Kalsi, Dehradun district. A trek goes to Har-Ki-Doon towards the northeast. Har-Ki-Doon is a wide and beautiful valley of the Tons River, where the GWS is located. Badasu valley, Purola, Rama valley, and Kamal valley are the beautiful natural locales. Another trek from Yamunotri goes

Fig. 6.2 Uttarkashi-Gangotri circuit: **a** Yamunotri temple and its environs, and **b** spectacular view of the Gangotri temple and its environs

to Harsil town towards the northeast, which lies on the way to Gangotri. Harsil is a nature's gift, a spectacular place, situated on the bank of the Bhagirathi River. It is an army base camp. To enter the Harsil valley, an inner line permit is required. From Uttarkashi, a road leads to the Gangotri pilgrimage. It is about 100 km long, along the Bhagirathi valley, which is declared as an 'eco-sensitive zone'. Some important natural locales are located between Uttarkashi and Gangotri. These places are Maneri, Bhatwari, Gangnani, Purga, Harsil, Mukhawa, and Bhaironghati. From Gangotri, a trek of about 16 km goes to Gaumukh, a source of the Bhagirathi River. A hydropower project is constructed in Maneri. The idol of Goddess Gangotri lies in the Gangotri temple in the pilgrimage season (summer season). In winter, the idol is shifted to the Mukhawa village, which is the winter home of the Goddess. From Bhaironghati, a deep gorge formed in the Bhagirathi River, a trek goes towards north up to the Tibet border. This circuit is very important because two pilgrimages—Yamunotri and Gangotri are situated on the banks of the Yamuna and the Ganga rivers. Har-Ki-Doon and Harsil are the two spectacular natural locales, located in this circuit.

6.4   Dehradun-Mussoorie-Chakrata-Rishikesh-Haridwar Circuit

This is the most important and world-famous tourist circuit. Dehradun, Mussoorie, and Chakrata are natural tourist places whereas Rishikesh and Haridwar are the two famous valley pilgrimages (Fig. 6.3). The city of Dehradun lies in the beautiful Doon valley, which stretches between the Ganga River in the east and the Yamuna River in the west. The RJNP surrounds it from two sides—east and south. Paonta Sahib is situated in the west and Mussoorie city lies in the north of Dehradun city. Dehradun is known for its aromatic basmati rice, educational institutions, and Army cantonments. It has several places of tourist interests. Mussoorie lies at an altitude of 1900 m (average), 30 km from Dehradun towards the north. It is a beautiful hill town, also known as the queen of the hills. Chakrata, a cantonment area, and a hill town is located at an altitude of 2200 m, 98 km from Dehradun. The town is surrounded by dense coniferous forests. Summers are very pleasant and during winters, Chakrata experiences heavy snowfalls. Haridwar, Rishikesh, and Dehradun have a triangle-shaped feature. Rishikesh 44 km from Dehradun is located on the bank of the Ganga River. After 24 km, Haridwar pilgrimage lies. From Dehradun, it is 54 km. A 24 km long belt between Rishikesh and Haridwar is known as the 'saffron belt' because many *Sadhus* (sages) live here and perform meditation. Rishikesh is known as the 'yoga capital of the world' and Hardwar/Haridwar is known as the 'Gateway to Lord Shiva' and 'Lord Vishnu'. About 60% of the total tourists/pilgrims of Uttarakhand Himalaya visit this circuit. The circuit is well connected by road, rail, and air transportation.

Fig. 6.3 a Rajpur Road, Dehradun, **b** a panoramic view of Mussoorie, **c** Haridwar pilgrimage, and **d** Rishikesh pilgrimage

Fig. 6.4 a Badrinath Dham on the right bank of the Alaknanda River, **b** Adi Badri temples, **c** Nandprayag, and **d** Karnprayag

6.5   Panch Badri Circuit

This is purely a cultural circuit that lies in the Chamoli district where mainly pilgrims visit. Lord Vishnu is the adored deity in this circuit. His idols are installed in five places—Vishal Badri (Badrinath) , Yogadhyan Badri, Bhavishya Badri, Vridha Badri, and Adi Badri (Fig. 6.4). Badrinath temple is located on the left bank of the Vishnu Ganga (the Alaknanda River) between Nar and Narayan mountain peaks, which is a sheet of Lord Vishnu. The temple is believed to be renovated by Adi Guru Shankaracharya in ninth century AD. The last village of India 'Mana', is located at a distance of three km from Badrinath. A trekking route to Satopanth Lake starts from the Mana village. Badrinath temple is opened for the devotees in the last week of April and closed in the last week of October. During winters, the idol of Lord Vishnu is shifted to Joshimath, a town and a winter abode of Lord Vishnu. Yogadhyan Badri is another form of Lord Vishnu located in the Pandukeshwar village, near Govind Ghat. Lord Vishnu resides in a meditation pose thus it is called Yogadhyan Badri. Yogadhyan Badri is considered as the winter abode of the festival-image of Badrinath. Bhavishya Badri is situated in the Subhain village after Tapovan, about 17 km from Joshimath town in the Dhauli Ganga basin. This place is predicted to be the future Badrinath as its name suggests. Lord Narsimha, an incarnation of Lord Vishnu, resides in Bhavishya Badri. A trail leads to the Kailash-Mansarovar, along the Dhauli Ganga, from here. The holy Vridha Badri temple is situated in the Animath Village, in the Dhauli Ganga basin, 7 km from Joshimath. It is believed that Lord Vishnu resides here in the form of an old man. In the religious wisdom of Hindus, it is said that sage Narada undertook penance here. The temple of Vridha Badri remains open throughout the year. Adi Badri, a group of temples, is situated in Adi Badri service center, about 19 km from Karnprayag, on the way to Gairsain, the summer capital of Uttarakhand. Ata Garh, a perennial tributary of the Pindar River, flows here. The temple is believed to be established by Adi Guru Shankaracharya. Besides five temples of Lord Vishnu in different river valleys—Vishnu Ganga, Dhauli Ganga, and the Pindar River—there are four Prayags (meeting place of two rivers)— Keshavprayag, Vishnuprayag, Nandprayag, and Karnprayag situated in this holy pilgrim circuit. Keshavprayag is situated on the confluence of the Vishnu Ganga (the Alaknanda River) and the Saraswati River in Mana. A natural stone bridge, called 'Bhim Pul', lies on the river Saraswati. In Vishnuprayag, two rivers—the Dhauli Ganga, originating from Niti Pass and the Alaknanda River, is originating from Bhagirathi Kharak glacier and Satopanth Lake—meets. Nandprayag is famous for the confluence of the two rivers—Nandakini (originating from a glacier below Nanda Ghunti in Chamoli district) and the Alaknanda River. The last Prayag in this circuit is Karnprayag, where the Pindar (originating from the Pindari Glacier in Bageshwar district) and the Alaknanda River meet. The entire circuit is one of the largest circuits in the Uttarakhand Himalaya. There are many other cultural and natural places situated in this circuit. One of the natural places is Auli, situated about 15 km from Joshimath in an upslope area. Auli is the best place for skiing, which is practiced during winters.

6.6   Panch Kedar Circuit

Panch Kedar circuit is devoted to Lord Shiva which includes Kedarnath, Tungnath, Rudranath, Madhyamaheshwar, and Kalpeshwar (Fig. 6.5). The temple of Kedar-nath is believed to be established by the Pandavas when they were on their way to heaven. It is believed that Adi Guru Shankaracharya attained salvation near the Kedarnath temple (Bhairav Jaap) after renovating it at the age of 32 years. All temples of Panch Kedar are situated in the serene sacred sites, close to the mighty Himalaya, which can be travelled by trekking only. This circuit is the biggest one in terms of trekking, which needs 10–12 days to be covered. Kedarnath is one of the famous pilgrimages of Panch Kedar, which can be trekked from Gaurikund, the last roadside settlement. It is a 16 km trekking route along the river Mandakini. A new trekking route has been constructed to reach Kedarnath Pilgrimage. The old trekking route, which passed through Rambada, was devastated in 2013 when cloudburst-triggered debris flow and flash flood occurred in the Kedarnath valley. Kedarnath temple is situated on a gentle slope in alpine pasture below Kedar peak, at an altitude of

Fig. 6.5 **a** Kedarnath temple and its environs, **b** Madhyamaheshwar temple, **c** Kalpeshwar temple, and **d** Rudranath rock temples

about 3500 m, on the bank of the Mandakini River, now in Rudraprayag district. The temple remains open between April and October. Tungnath, a temple of Lord Shiva, is situated at an altitude of 3600 m, approximately three km from Chopta, a service center, in Rudraprayag district. Tungnath area is an alpine pastureland, remains closed during winters because of heavy snowfall. It has a picturesque land-scape from where the mountain peaks of the Central Himalaya such as Nanda Devi, Kedarnath, Chaukhamba, and Neelkanth can be seen. Rudranath is a natural rock temple (2286 m), lying in Chamoli district, a 20 km trek from Gopeshwar. Thick forests of rhododendrons are found on the trekking route. People worship Neelkanth Mahadev. Several pools (Kunds) exist around the temple. They are Surya Kund, Chandra Kund, Tara Kund, and Mana Kund. Three peaks of Nanda Devi, Nanda Ghunti, and Trishul can be seen from Tungnath. Madhyamaheshwar lies at an alti-tude of 3200 m, in the lap of Mount Chaukhamba in Rudraprayag district. It can be reached from Mansona and Kalimath. A road goes up to Ransi village and from Ransi trekking route starts passing through Gaundar. The length of the trekking route is about 19 km. The charming valley of Madhyamaheshwar Ganga can be seen on the way. Varieties of wildlife can also be seen and forest landscape changes according to the change in altitude. About 2 km upslope from Madhyamaheshwar temple, Budha (old) Madhyamaheshwar is situated, which is an alpine pastureland.

Kalpeshwar temple is situated in Urgam valley of Chamoli district at an altitude of about 2200 m. Urgam valley is rich in dense forests, apple orchards, and cultivation of potatoes. Urgam can be reached from Helang (on the way to Badrinath) , and from Urgam, a 2 km old trek goes to Kalpeshwar. The Kalpganga flows in the Urgam valley, which meets the Alaknanda River at Helang. Two prayags are situated on the way to the Panch Kedar circuit. Rudraprayag is a town and cultural place where the Mandakini River meets with the Alaknanda River. The second Prayag in this circuit is Sonprayag, which is situated on the banks of the Son Ganga and the Mandakini River.

6.7 Devprayag-Kunjapuri-Chandrabadni-Surkanda-New Tehri Circuit

This circuit has three temples of Shakti (female deities), which are also called the 'Shakti Peeths', situated in Tehri district. Kunjapuri temple is situated in Narendra Nagar about 20 km from Rishikesh. Surkanda temple is situated between New Tehri and Dhanolti. A small trek of 1.5 km, starting from Kaddu Khal goes to the temple. Chandrabadni temple is situated between Srinagar and New Tehri road. From road head, a three km trek goes to the temple. All these temples lie above 2500 on the mountain peaks, which are covered by snow during four months of winter (Fig. 6.6). Devprayag is situated on the confluence of the Bhagirathi and the Alaknanda Rivers. All these places have cultural importance. New Tehri lies at an altitude of 2200 m (average) and is a planned town. Tehri high dam is situated on the Bhagirathi and Bhilangana rivers. Water sports have been started in the Tehri high dam reservoir. This circuit is important for both tourists and pilgrims. This circuit is known as the 'Shakti Peeth Circuit'.

6.8 Chilla-Kotdwar-Lansdowne-Pauri Circuit

This circuit is characterized by both natural and wildlife tourism. All the tourist places are located in the Pauri district. It starts at Chilla, a center for a wildlife safari. Chilla is situated in the dense forest area of RJNP, about 8 km from Haridwar. A 50 km area of the core region of the park can be visited in an open jeep. It takes about 2 hours to complete one round. A variety of wildlife can be seen during the safari. Kotdwar town is situated about 75 km from Chilla. One way goes through dense forests and others go through Najibabad. Kotdwar is situated in the foothills of the mountain known for the hermitage (Ashram) of sage Kanva, which was the birthplace of the king Bharat. Lansdowne, a hill town and a cantonment area, lies about 30 km from Kotdwar, on a hill slope at an altitude of about 1900 m. The landscape is spectacular. The Himalayan mountain ranges can be seen from the town. During

Fig. 6.6 a Devprayag, **b** Chandrabadni temple, **c** New Tehri town, and **d** Surkanda temple

summers, the climate is pleasant and in the winter season, the town receives several snow-spells. Forest landscape varies from pine forests in the low altitudes to mixed oak forests in the Middle Himalaya, and coniferous forests in the highlands. Pauri is an administrative town, situated on the northeast-facing slope at an altitude of about 1800 m. Tourists and pilgrims who visit the highland pilgrimage Badrinath also visit the tourist places of this circuit. There are several other places of tourists/pilgrims' interests in this circuit (Fig. 6.7).

Source: By Author.

6.9 Ranikhet-Almora-Jageshwar-Kausani-Baijnath-Bageshwar Circuit

Ranikhet, a natural tourist place, is located in the Almora district of the Kumaon Himalaya about 30 km from Almora. Situated on the hill slope of the Ramganga basin at an altitude of about 1800 m, the landscape of Ranikhet town is spectacular (Fig. 6.8). The town has a cantonment area and it is surrounded by dense pine forests. Almora is an administrative town, a natural tourist place, and a cultural town of the

Fig. 6.7 **a** Wildlife in a dense forest area of CWS, **b** spotted deers in an open area of CWS, **c** Kanva Ashram (hermitage) at Kotdwar, and **d** panoramic view of Pauri town

Fig. 6.8 **a** Magh Mela in Bageshwar, the Gomati River joining the Saryu River and **b** Ranikhet Township

Kumaon Himalaya. It is situated on a gentle hill slope in the Kosi River basin. The town is about 90 km from Kathgodam, the nearest railway station. About 30 km from Almora, a group of 124 temples of Jageshwar are situated on the left bank of the Jataganga River, a picturesque valley, which is covered by dense coniferous forests. It is a holy place and one of the pilgrimage sites. Three tourist places of this circuit—Ranikhet, Almora, and Jageshwar are located in Almora district and the other three—Kausani, Baijnath, and Bageshwar are located in the Bageshwar district. Kausani is a beautiful hill town situated at an altitude of about 1900 m, 30 km from Almora. The beautiful valley of Someshwar in the south and Baijnath town in the north can be seen from Kausani. The great Himalayan ranges with Mounts Trishul, Nanda Devi, and Chaukhamba can also be seen from here. It was the hometown of the late poet Sumitra Nandan Pant. Mahatma Gandhi visited Kausani during the freedom struggle of India. Baijnath, situated on the bank of the Gomti River, is a small town of Bageshwar district. The town is famous for its ancient temples of national importance. The temples were built by Narsingh Deo. Bageshwar, a cultural town famous for Shiva temple, is a holy town, situated on the confluence of Gomati and the Saryu River, about 20 km from Baijnath town (Sati 2017). Kapkot town lies about 20 km from Bageshwar, which leads to the Pindari glacier through trekking. Gwaldam and Kosi are two other places of tourists' interest, located in the proximity of the circuit.

6.10   Mukteshwar-Nainital-Bhimtal-CNP-Rudrapur Circuit

This circuit is located both in Nainital and USN districts. Mukteshwar is a cultural town, situated on a hilltop about 2,500 m altitude, 30 km from Almora (Fig. 6.9). Dense coniferous forests are found in the surroundings of Mukteshwar. A 40 km road goes to Bhowali and then in about 20 km distance, Kathgodam is located.

Fig. 6.9 **a** Surroundings of the Mukteshwar temple and **b** Bhimtal and its environs

Nainital district is bestowed with seven perennial lakes, and Nainital and Bhimtal being amongst them. Nainital is a beautiful natural tourist place, situated at an altitude of 1900 m, 30 km from Kathgodam a railway station. Nainital has numerous educational institutions, the Kumaon University being one of them, which are runs by the state government. The forest landscape is spectacular and forest types vary from pine in the lower altitudes, mixed-oak forests in the middle-higher altitudes, and coniferous forests in the high altitudes. Bhimtal town is situated on the surroundings of a perennial lake, about 15 km from Nainital town. CNP, one of the biggest national parks of India and a tiger reserve, is a center for wildlife tourism. Finally, Rudrapur is an administrative city, capital of USN. The famous GB Pant University of Agriculture and Technology is situated here. The city is well connected by air, rail, and road, and is situated about 40 km from Kathgodam.

6.11 Champawat-Pithoragarh-Dharchula-Munsiyari Circuit

This circuit is located in two districts—Champawat and Pithoragarh. Champawat is a cultural town and a town of temples situated at an altitude of 1615 m. Devidhura temple, situated on a hilltop, is close to Champawat town, with the Lohavati River flowing nearby (Fig. 6.10). Pithoragarh is an administrative town and a natural tourist place, also known as little Kashmir. It is situated at an altitude of 1514 m in a serene environment and spectacular landscape. The Ramganga River (E) flows in the western part of Pithoragarh. Dharchula is the westernmost town of Uttarakhand located in close vicinity of Nepal, 83 km from Pithoragarh. It is situated on the bank of the Kali River at an altitude of 915 m (average of the city) and surrounded by the Himalayan mountain peaks. Munsiyari, situated at an altitude of 2200 m, is a village in the Pithoragarh district. Munsiyari, which literary means 'snow place', is known as the 'Switzerland of India' and 'little Kashmir' because the surroundings of the Munsiyari village have spectacular landscapes. The snow-clad mountain peak Panchchuli can be seen from Munsiyari. It is a base camp for trekking to Milam, Ralam, and Namik glaciers.

6.12 Domestic Tourists/Pilgrims' Inflow in Tourism Circuits

Domestic tourists/pilgrims' inflow in tourism circuits was analyzed (Table 6.2). Data were gathered from each tourist/pilgrim center and summed up at the tourist circuit level. The time-series data of 2000 and 2018 were analyzed, and a change in domestic tourists/pilgrims' inflow was obtained. In 2000, the highest tourists/pilgrims' inflow was in TC 2 (Dehradun-Haridwar), which was 65.2%, followed by TC 3 (Panch

Fig. 6.10 **a** Munsiyari village, **b** Pithoragarh town, **c** Dharchula town, and **d** Devidhura fair in Champawat

Badri 7.55%) and TC 1 (Uttarkashi-Gangotri 6.2%). The lowest inflow was in TC 9 (Champawat-Munsiyari). Total tourists/pilgrims' inflow in all circuits were 9.205 million in 2000. In 2018, this number increased to 33.035 million. Of this, 74.67% inflow was in TC 2. In TC 5 (Devprayag-New Tehri), the inflow was 6.27% and it was 4.49% in TC 8 (Mukteshwar-Rudrapur). In terms of change in tourists/pilgrims' inflow, the highest increase was in TC 2, followed by TC 5, TC 8, and TC 9. A total of 258% increase in domestic tourists/pilgrims' inflow in all tourist circuits was observed during the period.

6.13 Foreign Tourists/Pilgrims' Inflow in Tourism Circuits

Foreign tourists/pilgrims' inflow in tourism circuits in 2000 and 2018 was analyzed (Table 6.3). Total tourists/pilgrims' inflow was 53,105, out of which, the highest inflow was 50.75% in TC 2 (Dehradun, Mussoorie, Chakrata, Rishikesh, and Haridwar Circuit), followed by TC 8 (Mukteshwar, Bhimtal, Nainital, CNP, and Rudrapur circuit). Other tourist circuits had less than 20% inflow. In 2018, the foreign tourists/pilgrims' inflow has increased to 150,695. Out of which, about 40.13% of flow was in TC 2, followed by TC 5 (Devprayag, Kunjapuri, Chandrabadni, New Tehri, and Surkanda). The other circuits had less than 15% inflow. The

Table 6.2 Domestic tourists/pilgrims' inflow in tourist circuits

Tourism circuits		Tourists'/Pilgrims' inflow		Change % (2000–2018)
		2000	2018	
TC 1	Uttarkashi-Yamunotri-Harsil-Gangotri	560,458 (6.2)	1,161,998 (3.52)	107
TC 2	Dehradun-Mussoorie-Chakrata-Rishikesh-Haridwar	6,001,783 (65.2)	24,666,622 (74.67)	7342
TC 3	Panch Badri: Badrinath, Yogadhyan Badri, Bridha Badri, Bhavishya Badri, and Adi Badri; Vishnuprayag, Nandprayag and Karnprayag	695,332 (7.55)	1,046,987 (3.17)	50.57
TC 4	Panch Kedar: Kedarnath-Tungnath-Rudranath-Kalpeshwar-Madhyamaheshwar; Rudraprayag and Sonprayag	542,370 (5.89)	1,004,087 (3.04)	85.13
TC 5	Devprayag-Kunjapuri-Chandrabadni-Surkanda-New Tehri	388,615 (4.22)	2,071,142 (6.27)	432
TC 6	Chilla-Kotdwar-Lansdowne-Pauri	312,548 (3.4)	909,361 (2.75)	190
TC 7	Ranikhet-Almora-Jageshwar-Kausani-Baijnath-Bageshwar	194,936 (2.12)	349,627 (1.1)	79.35
TC 8	Mukteshwar-Nainital-Bhimtal-Corbett-Rudrapur	411,139 (4.47)	1,483,084 (4.49)	260
TC 9	Champawat-Pithoragarh-Dharchula-Munsiyari	97,749 (1.1)	342,432 (1.04)	250
Total		9,204,930 (100)	33,035,340 (100)	258

Table 6.3 Tourists/pilgrims' inflow in tourism circuits (foreigners)

Tourism circuits	Tourists'/Pilgrims' inflow		Change % (2000–2018)
	2000	2018	
TC 1	1178 (2.22)	3082 (2.05)	162
TC 2	26,949 (50.75)	60,468 (40.13)	124
TC 3	–	1064 (0.71)	–
TC 4	–	1604 (1.1)	–
TC 5	7321 (13.79)	46,289 (30.7)	532
TC 6	310 (0.58)	12,775 (8.48)	4021
TC 7	5197 (9.79)	6851 (4.55)	31.82
TC 8	11,565 (21.78)	17,693 (11.74)	52.99
TC 9	585 (1.1)	869 (0.58)	48.55
Total	53,105 (100)	150,695 (100)	184

highest increase in tourists/pilgrims' inflow was noticed in TC 6 (Chilla, Kotdwar, Lansdowne, and Pauri), which was 4021%, followed by TC 5, TC 1 (Uttarkashi, Yamunotri, Harsil, and Gangotri), and TC 2. The other circuits had less than 60% increases. A total of 184% increase was registered during the period 2000–2018.

6.14 Conclusions

In this chapter, the major tourist/pilgrim circuits of the Uttarakhand Himalaya have been described as natural, cultural, park and wildlife, and adventure circuits separately or combined. Within the major tourists/pilgrims' circuits, some small circuits are also located. However, their descriptions have been given within the main tourist circuits. Out of the total tourists/pilgrims' circuits, few circuits have been developed where substantial tourism infrastructural facilities are available. Many of them are remotely located with inadequate infrastructural facilities. Some pilgrim circuits are only accessible through trekking. In the meantime, the potential for sustainable tourism development in these circuits is enormous. There are world-famous pilgrimages, tourist locales, and wildlife sanctuaries in the Uttarakhand Himalaya. The need is to develop them sustainably through providing transportation, accommodation, and institutional facilities.

References

Sati VP (2013) Tourism practices and approaches for its development in the Uttarakhand Himalaya, India. J Tourism Challenges Trends 6(1):97–112

Sati VP (2015) Pilgrimage tourism in mountain regions: socio-economic implications in the Garhwal Himalaya. South Asian J Tourism Heritage 8(1):164–182

Sati VP (2017) Cultural geography of Uttarakhand Himalaya. Today and Tomorrow, Printers and Publishers, New Delhi

Sati VP (2018) Carrying capacity analysis and destination development: a case study of Gangotri Tourists/Pilgrims' circuit in the Himalaya. Asia Pacific J Tourism Res 23(3):312–322. https://doi.org/10.1080/10941665.2018.1433220

Chapter 7
Case Studies of Important Tourism Routes

7.1 Introduction

The Uttarakhand Himalaya has plenty of tourists/pilgrims' routes. Many of them are accessible by roads, railways, and airways, but some are inaccessible, where the tourists/pilgrims carry out trekking. The author has divided the tourist/pilgrim centres into tourism circuits, and has described them in the sixth chapter. In this chapter, three tourism routes are described in details. These routes have multiple characteristics in terms of nature and types of tourism. Natural, cultural, adventure, and park and wildlife (ecotourism) tourism are practised in the places of tourists' interests in these routes. The author has selected these routes because they cover the three cultural realms of the Uttarakhand Himalaya. The first route covers the part of Chamoli district in the Garhwal Himalaya and parts of Almora and Nainital districts in the Kumaon Himalaya. The second route covers Dehradun and Uttarkashi districts and the third route covers Rudraprayag and Chamoli districts. The prominent tourist places of the Uttarakhand Himalaya are situated in these routes. These routes are characterised by too many and too less infrastructural facilities because of variation in terrain and accessibility of tourist places. Many of them are trekked by walking, mainly the highlands pilgrimages. Nainital, Almora, and Ranikhet, the natural locales of Nainital and Almora districts, Panch Kedar, the highlands pilgrimages of Rudraprayag and Chamoli Districts, and Hanol, the pilgrimage of Dehradun district; are the prominent tourism centres. The main objective of composing this chapter was to describe in details about the natural and cultural significance of the tourist places. Further, it aims at framing policies and planning for sustainable tourism development. The author travelled these routes and the description about them is based on his personal observations/experiences.

V. P. Sati, *Sustainable Tourism Development in the Himalaya:*
Constraints and Prospects, Environmental Science and Engineering,
https://doi.org/10.1007/978-3-030-58854-0_7

7.2 Karnprayag-Gwaldam-Almora-Nainital-Ranikhet-Gairsain Route

This journey begins from Karnprayag to Nainital through Gwaldam, Kausani, and Almora, and it comes back to Karnprayag through Ranikhet and Gairsain. Karnprayag is both a cultural and natural tourist place, situated on the confluence of the Alaknanda and the Pindar rivers. Tourists/pilgrims visit and stay at Karnprayag while on their way to Badrinath and other natural locales. It is believed that Karna, the son of Kunti, performed penance on the meeting place of these two rivers. A Karna Kunda and a temple of Karna are situated on a big rock close to these two rivers. The main road of Haridwar-Badrinath bifurcates from Karnprayag. One road goes to Badrinath along the Alaknanda River and the other goes to Almora along the Pindar River, which further bifurcates to Ranikhet. A small town and service centre is situated on the bank of the Pindar River, 7 km from Karnprayag, is called Simli. Simli is a flat land, river terrace. The main road bifurcates from Simli. One goes to Gwaldam along the Pindar River and the other goes to Gairsain, the summer capital of Uttarakhand. The route proceeds towards Gwaldam, which is about 45 km from Simli. The landscape is very fragile along the road therefore it remains blocked in many places due to landslides, mainly during the monsoon season. Further, the condition of the road is very bad. However, the landscape of the Pindar river valley is spectacular. On the way, many small service centres—Nauli, Bagoli, Nalgaon, Narayanbagar, Panti, Ming Gadhera, Harmani, Kulsari, and Tharali are situated on the river terraces. From Tharali, one road goes to Deval, a trekking base for Bedni Bugyal, Ali Bugyal, and Roop Kund, the other road goes to Gwaldam. The landscape between Tharali and Gwaldam is panoramic. Lolti, Thal, and Tal are the natural locales of tourism importance. Having dense temperate forests, the Himalayan ranges and snow-clad mountain peaks can be seen from these locales.

Gwaldam is a small town and a service centre, situated at an altitude of about 1950 m. It lies in the transitional zone of the two cultural realms—Garhwal and Kumaon. The town is surrounded by dense coniferous forest. The entire area is protected therefore its sprawl could not be possible for the last three decades. It has geographical importance as it divides the watersheds of the Pindar River basin in the west and the Gomti River basin in the east. One night halt is required in Gwaldam. The temperature during the winter season is very low. At night, it freezes. Further, chilled waves blowing from the Himalaya accentuate the severity of the cold. The Gwaldam town faces the snow-clad Mount Trishul. Therefore, the cold winds enter the Gwaldam town in the morning and evening time, which makes it look like the Himalaya moving into the town. The golden rays of sun fall on the mount Trishul (a trident shaped mountain peak) at early morning. The Mount Trishul symbolises Lord Shiva and thus, the folks worship it. Below the Himalaya Mountain, alpine meadows and then dense coniferous forests can be seen. The environment is clean. Morning walk for an hour in the dense forest area strengthens body and mind. Food and beverages are all locally made, and are tasty and healthy.

After spending few mourning hours in Gwaldam, one can proceed to Baijnath, a historical town of Bageshwar district in the Kumaon Himalaya, situated on the bank of the Gomti River, at an altitude of 1100 m. A group of 18 temples of Lord Shiva, Goddess Parvati, and other folk deities are situated on the river bank. These temples were built by Katyuri rulers during the thirteenth century (1202 AD) and King Gyan Chand restored and rebuilt these temples. Currently, the Archaeological Survey of India is looking after them. A road leads to Bageshwar and Kapkot from Baijnath. Bageshwar is situated in the stretch valley of the Gomti and Saryu rivers. These two rivers meet in Bageshwar town. Bageshwar town is a cultural place and the administrative headquarter of the district. A temple of Baghnath (Lord Shiva) is situated on the meeting point of two rivers, where 20 days Magh Mela is celebrated in January. The valley of Gomti is very fertile where paddy and wheat grow largely. The alluvial landscape is spectacular. Bageshwar was known as Danpur during the eighth century AD. It is the base camp for trekking to the Pindari glacier. Kapkot a service centre lies about 20 km from Bageshwar. The journey to Bageshwar is very enjoyable.

On the same day, a journey can be made back to Baijnath and Kausani. After 4 km from Baijnath, a small town and service centre Garur lies in the Katyur valley. Kausani is located on a mountain peak at an altitude of about 2000 m, at a distance of 10 km from Garur. Kausani has a picturesque landscape. It is the hometown of the eminent poet Sumitra Nandan Pant. A spiritual hermitage (Ashram) of Anashakti is situated here where Mahatma Gandhi spent some days. A long-range of the Himalaya Mountain can be seen from Kausani. Kausani town lies in the water parting line for the two watersheds. The Gomti River watershed lies in the west part and the Kosi River watershed lies in the east part. Forest landscape of Kausani is spectacular, which varies from pine in the lower altitude to mixed-oak and coniferous in the highlands. One night stay in Kausani is sufficient. The next morning, one can proceed to Almora. A beautiful and stretch valley of Someshwar lies in the Kosi River basin about 20 km from Kausani. The Someshwar valley is very fertile for growing various crops—grains, fruits, and vegetables. Kosi town is situated on the bank of the Kosi River about 10 km from Someshwar. Kosi is a beautiful valley where the 'Vivekanand Institute of Hill Agriculture Research' is situated on the bank of the Kosi River. The famous sun temple lies in the Katarnal village, which is located close to Kosi town in an upslope. The 'GB Pant Institute of Himalayan Environment and Development' (GBPIHED) is located in Katarmal. Almora is known as the cultural centre of the Kumaon Himalaya. It is equally important for its natural beauty. The Himalaya Mountain can be seen from Almora from three sides. The city is located at an altitude of about 1950 m. It has a range of accommodation—hotels and restaurants. Local food and beverages are served to tourists.

The route continues to Nainital through Mukteshwar and Bhawali. Mukteshwar, about 38 km from Almora, is situated on a hilltop at an altitude of 2500 m. It has a serene environment surrounded by dense coniferous forest. On the top of the hill, a temple of Lord Shiva lies. Pilgrims visit Mukteshwar and worship Lord Shiva. Mukteshwar is believed to be the place where pilgrims get rid of the cycle of birth and death as the Hindu scriptures say. Bhawali is situated on the way to Nainital, about

35.4 km from Mukteshwar. Ramgarh is located between Mukteshwar and Bhawali. Bhawali has a healthy climate. A tuberculosis hospital lies in Bhawali. The dense pine forest is found in and surroundings of the town. Nainital, a hill town/summer resort, is situated at an altitude of 2000 m on the banks of Naini Lake, about 12 km from Bhawali. It has temperate climate which acedes extremely cold during the winters and remains pleasant during the summers. At lower altitude, pine forests are found while in the surroundings of Nainital, mixed-oak and Deodar trees are found. Nainital is famous for the Mall Road and Naina Devi temple. Tourists walk and enjoy Mall road in morning and evening. On Nanda Ashthami, a fair is celebrated in the temple of Naina Devi. It continues for eight days. Further, fairs are celebrated on Navratra and Chaitra Sakranti. On the top of the town, Naina Peak, which is also known as China Peak, is situated. It can be reached by trekking 8.6 km from Mally Tal. From Naina Peak, a long-range of the snow-clad Himalaya is seen. For tourists two nights stay in Nainital town is enjoyable.

To come back, the same route can be followed up to nearby Almora town, from where a road bifurcates to Ranikhet. The entire landscape up to Ranikhet is awesome. Ranikhet is situated on a hill slope in the Ramganga River basin about 57 km from Nainital, surrounded by dense pine forest. The town has an army cantonment area and it is located at an altitude of about 1869 m. The town has a variety of accommodation—hotels and restaurants. Fresh and delicious traditional food is available at a reasonable cost. The Himalayan mountain ranges can be seen from the town. A night stay in Ranikhet town is recommended. Next morning, one can move to Karnprayag through Dwarahat, Chaukhutia, and Gairsain. Dwarahat, known as 'Old Dwarika', is situated in the stretch valley of the Ramganga River. Chaukhutiya, a town and service centre, lies on the bank of Ramganga, about 19 km from Dwarahat. From Chaukhutiya, a road bifurcates to Ramnagar and other goes to Karnprayag. About 34 km from Chaukhutiya, Gairsain, a small town and the summer capital of Uttarakhand lies on the foothills of the Dudhatoli Mountain range. Gairsain topography is a flat alluvial plain, very fertile for growing various crops. Ramganga flows through the town. Diwalikhal is a small village, situated on a hill ridge at an altitude of about 2200 m, eight km from Gairsain. Five km from Diwalikhal, Bhararisen village is situated on a beautiful ridge, where a dairy farm and the Uttarakhand legislative assembly house are located. On the way to Karnprayag, Adi Badri pilgrimage is situated on the bank of Atagarh, is one of the Panch Badri. After 19 km from Adi Badri, Karnprayag is located. The road goes through Simli village/service centre.

The entire route of about 381 km has spectacular landscapes. The road traverses through several river valleys topographies—gorges, river terraces, stretch alluvial plains, middle altitudes, and the highlands, parallel with the snow-clad Himalayan ranges. Forest landscape, climate, and food and beverages are excellent on the entire route. This tourism route has many natural and cultural locales. The culture and geo-environment along the route vary from the Garhwal Himalaya to the Kumaon Himalaya, the two different geo-cultural entities (Sati 2017). Further, it has sub-cultural realms in both the regions. The route can be travelled within seven days. Figure 7.1 shows the distance between tourism centres and their altitudes. Figure 7.2 shows the topographies of four major tourism sites.

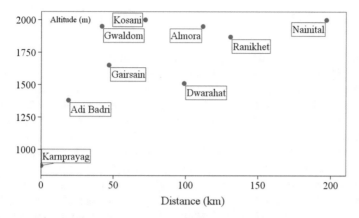

Fig. 7.1 Major tourist places of Karnprayag-Nainital tourist route: distance and altitude

Fig. 7.2 a Gwaldam town, Mount Trishul is seen in the background **b** Baijnath temple, the Himalayan mountain range is seen in the background **c** Sun temples in Katarmal, Almora, and **d** Naini Lake in Nainital. *Photo* By author

7.3   Kalsi-Chakrata-Tiuni-Hanol-Mori-Purola Route

This tourist route covers the places of Jaunsar-Bawar (Chakrata sub-administrative division) in Dehradun district and the Rawain valley of Uttarkashi district. The major tourist places are Kalsi, Chakrata, Tiuni, Hanol, and Purola. The journey begins from Dehradun. On the way, Kalsi, a small town and a cantonment area, is situated on the bank of the Yamuna River about 45 km from Dehradun. Its altitude is about 780 m. Dakpatthar, a hydropower project; constructed on the Yamuna River is situated close to Kalsi town. Its heritage is considered important as the famous Ashoka rock edict is located here, which is written in Pali language. The town is situated on the foothills. Mountains start from here. Sahiya is a village and a service centre situated about 19 km from Kalsi. Chakrata is the next destination, 25 km from Sahiya. Chakrata town is situated on a ridge at an altitude of 2200 m, known for its pollution-free atmosphere and serene environment. From Dehradun, it is at a distance of about 98 km. It has a cantonment area. The landscape is spectacular. Temperate forests—mixed-oak and coniferous forests make the forest landscape beautiful. About 20 km downstream, a Tiger Fall is situated, which is a major tourist attraction. It has a pleasant climate during the summers and during winter, several spells of snowfall. Tourists' inflow is higher than the accommodation facilities because the town is small and few hotel facilities are available. Traditional food and beverages are famous, which tourists enjoy during their visit to Chakrata. A night stay is recommended. A road goes ups and downs up to Tiuni village, a service centre, which is situated on the bank of the Tons River (also known as the Tamas River). The altitude goes up to 2700 m on the way to Tiuni. Temperate forests, mainly deodar are found above 2000 m altitude. The roadblocks during the two months of the winter season—December and January because of heavy snowfall. Further, because of fragile landscape and heavy rainfall during the monsoon season, landslides occur along the road, causing roadblocks.

Tiuni village is located in the core region of Jaunsar-Bawar at an altitude of 1050 m. It is also the border area between Dehradun district of Uttarakhand and Sirmour district of Himachal Pradesh. A village called Gorchha, lying close to Tiuni village has an ancient cave, which is the longest—7 km long and 15–20 feet high, which goes to the deep interior of the earth. People believe that the cave is the oldest one, existing from the time of Pandavas. A night stay in Tiuni village can refresh tourists/pilgrims. Another important place in this route is Hanol village, which is 15 km from Tiuni. The Hanol village is located on the bank of the Tons River at an altitude of 1050 m. The word Hanol is derived after the Brahmin 'Huna Bhatt'. The village is surrounded by lush green forests. Hanol is famous for the Mahasu Devta (Maha Shiva) Temple. The temple was built in the ninth century. The architect of the temple belongs to Huna style. The Archaeological Survey of India is taking care of the temple. The Mahasu Devta fair is celebrated in August every year. Mori is situated on the bank of the Tons River, at an altitude of 1150 m, about 16 km from Hanol, known as the sleepy hamlet. It has a panoramic/scenic beauty, surrounded by paddy fields. Mori village has a unique culture and history. The people believe that they are the descendants of the Kauravas and Pandavas. In the surroundings of

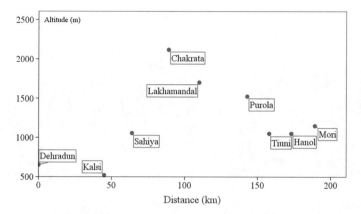

Fig. 7.3 Major tourists places in Kalsi-Chakrata-Tiuni-Purola route: distance and altitude

Mori, dense pine forests are found. An ancient temple of Duryodhana is the major tourist/pilgrim attraction. Natwar is located close to Mori, where a temple of Karna is located. The temple has a rectangular wooden structure. Water sports (river rafting) are played in the Tons River and the best campsite is Mori village.

Purola is a village and also the administrative headquarters of Purola taluk, situated in the famous Rawain valley of Uttarkashi district. Located at an altitude of 1524 m, Purola is 32 km from Mori. The Rawain valley is wide, fertile, and beautiful. It has a spectacular landscape. The valley is surrounded by hills and mountains. The Kamal River flows in the valley and divides the watershed of the Tons River. Two days halt at Purola is sufficient to visit the entire Rawain valley. The road distance from Purola to Kalsi is 102 km. Lakhamandal, Nainbag, and Lakhwar are situated along the Yamuna River between Purola and Kalsi. This tourism route has three cultural realms—Garhwali, Jaunsari, and Rawain. All three realms have their own importance. The landscapes also vary from one to other. Five days are required to travel this route. Figure 7.3 shows the distance between the tourists' places/pilgrimages and their altitudes. Figure 7.4 shows topography of four important tourist/pilgrim centres.

7.4 Rudraprayag-Kedarnath-Madhyamaheshwar-Tungnath-Rudranath-Kalpeshwar Route

This route has four types of tourism—cultural, natural, adventure, and wildlife and it starts from Rudraprayag. Kund is a village situated on the bank of the Mandakini River about 30 km from Rudraprayag. It lies at the junction of four routes—Kedarnath, Madhyamaheshwar, Chopta, and Rudraprayag. A route goes to Kedarnath along the Mandakini River valley. A motor road goes up to Gaurikund, which is 38 km from Kund. On the way, Guptakashi, Fata, and Sonprayag are situated. From Gaurikund, a 16 km trekking route goes to Kedarnath. Before 2013, the trekking route was

Fig. 7.4 a Aerial view of Dehradun valley, **b** Way to Tiuni from Chakrata, **c** Tiger Fall in Chakrata, **d** Picturesque landscape of a village on the way to Tiuni from Chakrata. *Photo* By author

14 km long through Rambada however, due to the tragedy that hit in Kedarnath 2013, the entire trek devastated and a new route of 16 km was constructed. The route goes through steep and precipitous slopes and glaciers, and snow-clad mountain peaks. Suitable accommodation facilities are available in Kedarnath, making a night halt possible. Kedarnath pilgrimage is situated on the bank of the Mandakini River on a gentle slope (Sati 2015). The surrounding areas of Kedarnath are snow-clad mountain peaks. It has a helipad, which serves pilgrims during the pilgrimage season. Chaurabari glacier lies 7 km upward from Kedarnath, which is the source of the Mandakini River. The next destination is Madhyamaheshwar. From Kedarnath, one has to come back to Kund. From Kund, there are two roads that lead Madhyamaheshwar. One goes through Mansona and the other goes through Raunlek and Ransi. Both roads lead to Gaundar, the last village, situated on the mountains' root in the Madhyamaheshwar valley. From Gaundar, Madhyamaheshwar is a 19 km trek, which passes through steep slopes, dense forests, and rough and rugged terrain. Madhyamaheshwar is situated on the lap of the Mount Chaukhamba, left side of the Madhyamaheshwar River. It is an alpine pastureland (Bugyal). About 2 km upslope, Budha (old) Madhyamaheshwar is situated. The entire pastureland is full of flowers and medicinal plants, which blossom in September and October, considered to be the

main pilgrimage season. A small Dharmashala is available for night halt. The nature lovers trek Madhyamaheshwar with tents and food material, and they further trek to Mount Chaukhamba. After trekking Madhyamaheshwar, the pilgrims come back to Ukhimath, which lies in the close vicinity of the Kund. Ukhimath is the winter abode of Lord Kedar. The idols of Lord Kedar from Kedarnath and Madhyamaheshwar are brought to Ukhimath for the winter season. The trek goes to Chopta, about 45 km from Ukhimath. Chopta lies in a Bugyal. In the lower altitudes, coniferous forests are found. The panoramic view of the Himalayan mountain ranges with the Mount Chaukhamba can be seen in the proximity. It is a small service centre, known as the 'Switzerland of Chamoli district'. About three km upslope of Chopta, Tungnath temple is situated on the top of a mountain peak in a Bugyal, with a gentle slope. Tungnath is one of the Panch Kedars. The entire region receives heavy snow during the winters. It needs a maximum of four hours to make a complete visit of the pilgrimage and travel can be continued up to Gopeshwar town on the same day. From Chopta to Gopeshwar, the entire route goes through NDBR, which is known for rich biodiversity. The endangered musk dear is found in this sanctuary. The route is adventurous with dense forest cover and wild animals. About 7 km before Gopeshwar, a small village named Mandal is situated, where the famous temple of Anasuya Mata and Atri Muni hermitage is situated. Married couples worship Mata Anasuya for blessing them with a male child. Gopeshwar is an administrative town and a district headquarter, situated on hill slope, 8 km from Chamoli town, connecting Haridwar and Badrinath, along the Alaknanda River. The famous Gopinath temple and a tuberculosis hospital lie in Gopeshwar. The next destination is Rudranath, a group of rock temples, situated at about a 14 km trekking from Gopeshwar. The pilgrims can come back from Rudranath on the same day. After a halt at Gopeshwar overnight, the route goes to Kalpeshwar temple through Chamoli town and Urgam village along the Alaknanda River valley. From Urgam, a 2 km trek goes to Kalpeshwar temple. Finally, the pilgrims come back to Rudraprayag via Nandprayag, situated on the bank of the Nandakini and the Alaknanda River; Karnprayag situated on the bank of the Pindar and the Alaknanda River; and Gauchar, a service center, lying at a distance of 10 km from Karnprayag. The route is mainly devoted to pilgrimage tourism, known as Panch Kedar. Figure 7.5 shows the distance of tourist places and pilgrimages and their altitudes. Figure 7.6 shows the snow-clad surrounding areas of the Tungnath temple, and the Mandakini River meeting the Alaknanda River at Rudraprayag.

7.5   Conclusions

In this chapter, three tourism routes and their geographical and cultural importance have been described. These routes have a high potential for sustainable tourism development. In the meantime, some of the tourism locales on these routes are underdeveloped. Infrastructural facilities are lagging. Accessibility of some tourist places is very difficult, mainly of the highland pilgrimages. Accommodation and institutional facilities are not up to the mark. Therefore, it has been noticed that the tourists/pilgrims'

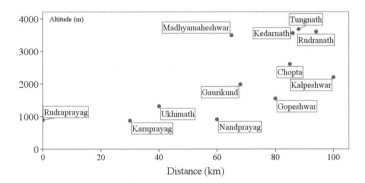

Fig. **7.5** Rudraprayag-Kedarnath-Madhyamaheshwar-Tungnath-Kalpeshwar   tourism   route:
distance and altitude

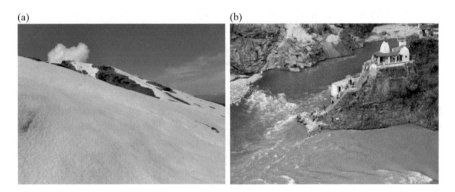

Fig. 7.6 **a** Tungnath covered by snow in Feb 2020, **b** Rudraprayag—the confluence of Mandakini
River (left) and Alaknanda River

inflow in these locales are considerably less whereas the geographical and cultural
bases of tourism are significantly high. It is suggested that the development of infrastructural facilities—transportation, accommodation, and institutional—will provide
ample opportunities to tourists/pilgrims to visit these routes, leading to sustainable
tourism development.

References

Sati VP (2015) Pilgrimage tourism in mountain regions: socio-economic implications in the Garhwal
    Himalaya. South Asian J Tourism Heritage 8(1):164–182
Sati VP (2017) Cultural geography of Uttarakhand Himalaya. Today and Tomorrow, Printers and
    Publishers, New Delhi

Chapter 8
Infrastructural Facilities for Tourism Development

8.1 Introduction

Infrastructure facilities include physical, institutional, and human resources that provide a support system for tourism development. Physical infrastructure is an important component of overall tourism development (Khadaroo and Seetanah 2008), along with accommodation and institutional facilities. Tourists demand a sustainable infrastructural facility so that they can feel homely when they are practicing tourism (Murphy et al. 2000; Crouch and Ritchie 2000). Infrastructural facilities are mainly related to the public sector and therefore the government role is noteworthy (Cooper et al. 2008; McConnell, 1985). Dwyer et al. 2010 stated that defining tourism infrastructure is not easy because it is not a single industry. Hansen (1965) and Mera (1973) defined that infrastructural facilities are the total of economic and social capitals such as roads, streets, bridges, public health and education. Ritchie and Crouch (2005) have defined it as public safety, transportation services, medical systems, financial systems and education systems. Tourism infrastructure is related to all components that are available in the destination to promote tourism (Swarbrooke and Horner 2001; Lohmann and Netto 2017) although most of the tourism infrastructure is used by the residents (Fourie and Santana 2011; Hadzik and Gabara 2014). Tourism infrastructural facilities are always related to sustainable tourism development (Heath 1992; UNWTO 2007; Sharpley 2009; Getz 1992; Formica and Uysal 1996; Garay and Canoves 2011). These facilities also influence the competitiveness of tourism destinations (Crouch and Ritchie 1999; Murphy et al. 2000; Sakai 2006). There has been established a strong relationship between tourism development and infrastructural facilities theoretically by many authors (Adobayo and Iweka 2014). The purpose of these facilities is to fulfill the desires of tourists/pilgrims (Popesku 2011). Tourism infrastructure attracts tourists/pilgrims through supplying transport, social and environmental infrastructure (Tourism and Transport Forum (TTF) 2012; Lovelock and Lovelock 2013).

V. P. Sati, *Sustainable Tourism Development in the Himalaya: Constraints and Prospects*, Environmental Science and Engineering, https://doi.org/10.1007/978-3-030-58854-0_8

95

The infrastructural facility is the key for tourism development mainly in the mountainous areas where the landscape is fragile and most of the tourist places are remotely located. Infrastructural facilities mainly include—accommodation, transportation and institutional support including medical and institutional ones. The Uttarakhand Himalaya is bestowed with numerous tourist places and pilgrim centers. Meanwhile, many of them are not accessible comfortably because of the lacking infrastructural facilities. The road condition is not good. Further, in many highland tourist places and pilgrimages roads are not available. Railways are limited to plain areas—Dehradun, Haridwar, Kotdwar, Haldwani, Kathgodam, and Rudrapur, which share only 7% of the total geographical area of the State. Helicopter facility, operating from Fata, is available only to the Kedarnath pilgrimages. There are only five airports in the state. One is in Dehradun, the state capital and the others are in Rudrapur, Pithoragarh, Chinyalisaur, and Gauchar. Recently, helicopter services have begun in Pithoragarh and Uttarkashi. However these services are not regular.

Accommodation is an important infrastructural facility for tourism development (Nest & Wings 2010). Tourism requires a range of accommodations that is suitable to the tourists/pilgrims pockets. In Uttarakhand, star hotels are available only in few of the major cities. In many tourist places and pilgrimages, which are located remotely, accommodation facilities are not adequate. The tourists/pilgrims' inflow is higher than what the accommodation facilities can support and therefore, in many highland tourist places and pilgrimages, the tourists/pilgrims plan their return on the same day (Sati 2018). The number of pilgrims is always higher than tourists in Uttarakhand because of the world-famous river valleys and highlands pilgrim centers. The economic level of the pilgrims is comparatively low; therefore they face problems related to reasonable accommodation.

Institutional facilities contribute a greater role in the tourism development. It needs human resource development, providing services and publicity of the tourist places and pilgrimages, which are yet to be explored and developed in Uttarakhand. Institutional facilities should be developed at the local level. At present, UTDB is a nodal agency of the state government, which plans and coordinates tourism development. Two agencies of the state government at divisional level—GMVN at Dehradun and KMVN at Nainital are providing basic infrastructural facilities to tourism development. Every district of the state is having a tourism department at its headquarters. However, the institutional facilities are significantly inadequate because of the inaccessibility and remoteness of the tourism places and pilgrim centers.

This chapter examines the infrastructural facilities—transportation, accommodation and institutional at the state, districts and local levels. All types of transportation are elaborated. Accommodation available at the tourism places and pilgrimages are described and institutional facilities for promotion of tourism are illustrated (Fig. 8.1).

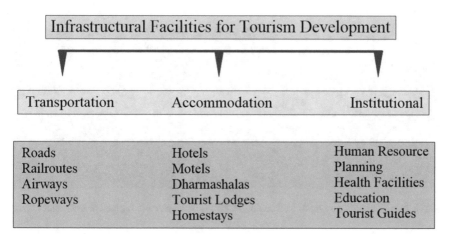

Fig. 8.1 Infrastructural facilities for tourism development. *Source* By author

8.2 Transportation Facilities

In the Uttarakhand Himalaya, transportation includes airways, railways, and roads. In most of the tourist places and pilgrimages, only road transportation is available. Further, road condition is not good because of the highly fragile slopes and difficult terrain. During the monsoon period, the roadblock is a common phenomenon because of heavy downpour and overflow of rivers and streams. Several highland tourist places and pilgrimages are having minimal road transportation. The tourists and pilgrims travel in the tourism places by trekking routes. The major means of transportation are elaborated as follows:

8.3 Airways

Uttarakhand has a high potential for expanding the air transportation because of its remoteness on the one hand and the other, it has numerous natural and cultural places. Further, the construction of roads and rail routes is not much feasible because of the high fragility of the landscape; airways may be a suitable means of transportation. Uttarakhand has five airports. India's leading airport is located in Dehradun, which operates direct flights to almost all the major cities of India. A new International Airport is under construction beside the domestic airport. Dehradun airport is strategically important because of its location is close to Tibet. Tourism to the Garhwal region in the forms of pilgrimage, natural and the adventure begins from Haridwar, which is located close to Dehradun thus the airport in Dehradun, has tremendous importance. Similarly, the Rudrapur (Pant Nagar) airport, located in USN, is an important airport for the Kumaon region, because the Rudrapur is known as the

(a)   (b)

Fig. 8.2 a Chopper is flying to Kedarnath, **b** Chopper has landed in helipad at Kedarnath. *Photo* By author

gateway to Kumaon region. The third airport is situated in Bharkot (Chinyalisaur) in the Uttarkashi district. The airport serves the strategic and operational requirements of the Indian Air Force. It is also important because the two pilgrimages—Yamunotri and Gangotri are located in its vicinity. Further, the airport has a strategic location owing to its proximity to Tibet. Gauchar airport is located in Gauchar service center, on the way to Badrinath, 10 km before Karnprayag in Chamoli district. The pilgrims visiting Badrinath and Kedarnath can use this airport. Presently, the airport is under army control. The fifth airport Naini Saini is located in Pithoragarh town in the Pithoragarh district. The airport was constructed in 1991, one of the strategically important airports of Uttarakhand. All these five airports are connected from Dehradun. They are operational throughout the year. Besides, a helipad is located in Fata, Guptakashi in Rudraprayag district, which is used for transporting pilgrims to Kedarnath pilgrimage and connecting to Dehradun (Fig. 8.2).

8.4 Railways

The state of Uttarakhand is mountainous. The construction of railways is not feasible. Railways connect only the towns/cities of the plain regions whereas mountainous mainland is not traversed by rail route. There are a total of 12 railway stations out of which, six are in the Garhwal region and six are in the Kumaon region. The main railway stations in the state are Dehradun, Raiwala, Haridwar, Rishikesh, Kotdwar, and Roorkee in the Garhwal region and Ramnagar, Tanakpur, Rudrapur, Lal Kuan, Haldwani, and Kathgodam in the Kumaon region. Dehradun, Haridwar, and Kathgodam are the big railway junctions, which are connected to all the major railway junctions of India. A railway line of about 100 km is under construction between Rishikesh and Karnprayag. This is India's most ambitious project which will be a great help for the pilgrims who are visiting the Badrinath and Kedarnath pilgrim

centers. In 2017, another railway project, which will connect the four pilgrimages—
Yamunotri, Gangotri, Kedarnath and Badrinath, has been launched.

8.5   Roadways

Roads are the major means of transportation in Uttarakhand. There are a total of
seven types of roads traversed in the state, which are national highways, state highways, major district roads, other district roads, village roads, light vehicle roads,
and bridle roads/border tracks. Time series data—2000 and 2019 were gathered
from the Public Work Department, Dehradun. Data shows that the total length of
roads was 17,571.8 km in 2000, which has increased to 41,374.96 km (135.47%
increase). The longest road was under village road—7446.23 km (42.38%) in 2000
and 23,953.59Km (75.89%) in 2019 (Table 8.1). The length of the National Highways was only 2.99% in 2000 and it increased to 5.05% of the total length in 2019.
The length of the state highways was 7.03% in 2000 and 10.92% in 2019. The
construction of roads is continuing in the Uttarakhand Himalaya and its length is
increasing. Except for border tracks (bridle roads), which has decreased by 9.82%
other types of roads increased substantially. A total of 135.47% road length was
increased during the last 19 years. Along with the major roads, there are a total of

Table 8.1 Category and length of roads in 2000 and 2019	Category of road	Length of roads (Km)		Change (%) (2000–2019)
		2000	2019	
	National highway	526 (2.99)	2091.34 (5.05)	297.59
	State highway	1235.04 (7.03)	4516.91 (10.92)	265.73
	Major district road	1364.15 (7.76)	2113.17 (5.12)	54.91
	Other district road	2714.61 (15.45)	4583.01 (11.1)	68.83
	Village road	7446.23 (42.38)	23,953.59 (75.89)	221.69
	Light vehicle road	315.77 (1.8)	536.69 (1.29)	69.96
	Border tracks	3970 (22.59)	3580.25 (8.65)	−9.82
	Total	17,571.8 (100)	41,374.96 (100)	135.46

Source Public Work Department, Dehradun, Uttarakhand 2020
Note Figures in parenthesizes are the percentage

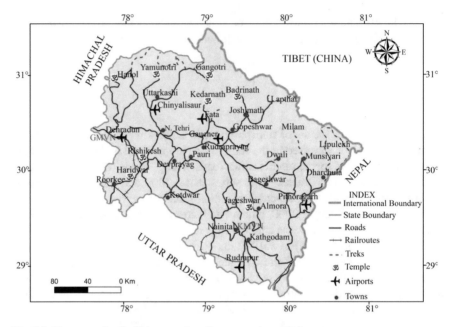

Fig. 8.3 Transportation facilities—roads, rail routes, treks, and airports

1000 bridges constructed on the rivers. Figure 8.3 shows transportation facilities and Table 8.1 shows category and length of roads in the Uttarakhand Himalaya.

8.6 Major National Highways

The state of Uttarakhand is connected by several National Highways (NHs). Here, only the National Highways, which are traversed within the state, are described. All the NHs are given a new number by the Road Transportation Department, Government of India. The NH No. 334 is connecting the two pilgrim centers—the valley pilgrimage at Haridwar and the highland pilgrimage at Badrinath with Delhi. It goes up to Mana village, which is situated on the confluence of the Saraswati River and the Alaknanda River, 3 km from Badrinath. It has a total length of 380.8 km. The NH No. 7 is connecting Ambala and Ponta Sahib with Dehradun, Rishikesh, and Haridwar with its total length of 92.8 km. Rampur, Rudrapur, Haldwani, and Nainital is connected by the NH No. 109. Its total length is 105 km. The NH No. 134 is connecting Rishikesh, Dharasu, Barkot, and Phoolchatti with its total length of 236 km. Dharasu-Gangotri NH No. 34 is 124 km. Rudraprayag-Gaurikund is located in NH No. 107 and the total length is about 76 km. Vikasnagar-Kalsi-Barkot towns are connected by NH No. 507 and the total length of the highway is 111 km. The NH No. 534 is connecting Bijnor, Najibabad, Kotdwar, and Pauri towns with a total length of 137 km. The NH No. 309 is connecting the two districts of Kumaon and Garhwal regions. The major towns and cities on the NH are Ramnagar, Dhumakot, Thalisain, Tripalisain, and Pabo. Its total length is 364 km. NH No. 109 is connecting Bhawali and Karnprayag through Almora-Ranikhet, Dwarahat, Chaukhutia, Gairsain, and Adi Badri. Its total distance is 235 km. Khatima-Tanakpur-Pithoragarh lies on the NH No. 9 and its length is 202 km. Gopeshwar-Okhimath NH No 107A is 84.2 km long passing through NDBR. On the way, Mandal, Chopta, Pothibasa, and Duggalbita service centers are located. Gangolihat-Berinag—Chaukori-Kanda-Bageshwar-Takula-Almora towns are connected by NH No. 309A. Its total length is 208 km. NH No. 9 is connecting Pithoragarh and Askot with a total length of 53 km. Almora and Panar are connected by the NH No. 309B and the total length is 82 km. The last NH in Uttarakhand is 707A, which connects Tiuni-Chakrata-Mussoorie-Chamba-New Tehri-Maletha with a total length of 310 km. The total length of NHs in Uttarakhand is 2954.1 km. All these NHs are under different agencies such as PWD, NHAI, BRO, and NHIDCL. The highest length of NHs is under PWD, which is 2091.34 km, followed by NHAI (379.96 km), BRO (382.8 km), and NHIDCL (100 km).

All the above-mentioned NHs are connecting the major tourist places and pilgrim centers of the river valleys and middle-altitudes. However, most of the highland tourists' places and pilgrimages are inaccessible where roads are not traversed. Further, the roads' quality is poor and landslides along the roads are common. This leads to road blockage and major accidents mainly during the monsoon season.

8.7 All-Weather Roads

Construction of all-weather roads, connecting the four highland pilgrimages with Rishikesh, is the ambitious project of the central and state governments. In 2014, when the BJP led Central Government resumed the power under the Prime Minister Shri Narendra Damodar Modi, the all-weather road project was started. The envisaged total length of all-weather roads is 889 km. It starts from Rishikesh and traversing along first the Ganga and then the Alaknanda River valley—Devprayag, Rudraprayag, Karnprayag, Nandprayag, Vishnuprayag, Pandukeshwar, and finally reaching the Badrinath pilgrim area. The second all-weather road starts from Rudraprayag, passes through Augustyamuni and Guptakashi, along the Mandakini River valley and it lasts to Gaurikund, the base camp of the Kedarnath pilgrimage. From Rishikesh, another all-weather road starts and it ends in Yamunotri and Gangotri, the two highland pilgrim centers. It passes through Narendra Nagar, Chamba, and then along the Bhagirathi River valley and the Yamuna River valley. Another all-weather road is being constructed in the Kumaon region, which is connecting Pithoragarh to Tanakpur town. All-weather roads are expected to be completed by 2022 and they will be the milestone for tourism development in Uttarakhand.

8.8   Accommodation

Accommodation is one of the major components/elements of tourism development. The Uttarakhand Himalaya is a developing state where about 50% of GSDP is contributed by tourism industry. However, substantial accommodation facilities are lagging. Uttarakhand needs a range of accommodation, which will serve the high-income level to the low-income level tourists/pilgrims. As Uttarakhand has many pilgrimages and cultural places thus the number of pilgrims is higher than the number of tourists. Further, the income level of pilgrims is considerably low to support their travel. Therefore, pilgrims need affordable accommodation facilities. The accommodation facilities available for tourism development in the Uttarakhand Himalaya are as follows:

8.9   Accommodation Units in the Major Natural and Cultural Locales

Table 8.2 shows accommodation units in the major natural and cultural places. First, the total number of hotels in the city/town level is analyzed. Indices are given and based on the indices; the towns/cities are divided into levels. The total

Table 8.2   Accommodation units in the major natural and cultural places

Indices	Levels	Towns/cities
Total hotels (1768)		
>100	High	Haridwar, Mussoorie, Nainital, and Dehradun
51–100	Medium	Yamunotri, Uttarkashi, Rishikesh, and Gangotri
<50	Low	Joshimath, Almora, Badrinath, Pauri, Ranikhet, Kausani, Kotdwar, Pithoragarh, CNP, and Bageshwar
Total rooms (31,065)		
>2000	High	Haridwar, Nainital, Mussoorie, and Dehradun
1000–2000	Medium	Rishikesh, Uttarkashi
<1000	Low	Yamunotri, Gangotri, Almora, Joshimath, Badrinath, Pithoragarh, Ranikhet, Kausani, Pauri, Kotdwar, CNP, and Bageshwar
Total beds (67,973)		
>5000	High	Haridwar, Nainital, Mussoorie, and Dehradun
1000–5000	Medium	Uttarkashi, Rishikesh, Yamunotri, Badrinath, Gangotri, Joshimath, and Almora
<1000	Low	Pithoragarh, Ranikhet, Pauri, Kausani, Kotdwar, CNP, and Bageshwar

Source UTDB and analyzed by the author

Fig. 8.4 Accommodation—hotels, rooms, and beds

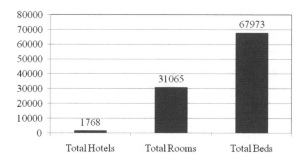

number of hotels in the selected tourist places is 1768 and all these hotels are registered with the tourism department. From the table, it is clear that Haridwar has the highest number of hotels (828), rooms (17,546), and beds (36,252), which is about 50% of the total accommodation. On the other hand, Bageshwar has the lowest number of hotels (4), rooms (73), and beds (135), about 0.23% of the total accommodation. The number of hotels > 100 (high level) are available in Haridwar, Mussoorie, Nainital and Dehradun. There are four cities/towns where the number of hotels is between 51 and 100 and these places are Yamunotri, Uttarkashi, Rishikesh and Gangotri. Many tourists' places have less than 50 hotels and they are Joshimath, Almora, Badrinath, Pauri, Ranikhet, Kausani, Kotdwar, Pithoragarh, CNP and Bageshwar. The total rooms in these hotels are 31,065. Haridwar, Nainital, Mussoorie, and Dehradun cities have more than 2000 rooms each, followed by Rishikesh and Uttarkashi. These two cities/towns have 1000–2000 rooms each available for tourists/pilgrims. Rooms below 1000 each are available in 12 selected natural and cultural places, which are Yamunotri, Gangotri, Almora, Joshimath, Badrinath, Pithoragarh, Ranikhet, Kausani, Pauri, Kotdwar, CNP, and Bageshwar. In all tourist places, a total of 67,973 beds are available. The highest beds above 5000 each are available in Haridwar, Nainital, Mussoorie, and Dehradun. From 1000 to 5000 beds are available in Uttarkashi, Rishikesh, Yamunotri, Badrinath, Gangotri, Joshimath and Almora. Rooms less than 1000 are available in Pithoragarh, Ranikhet, Pauri, Kausani, Kotdwar, CNP and Bageshwar. Besides these accommodations, from 2015 homestay facilities have been started in several places of Uttarakhand. A detailed description of homestay facilities is given in Chap. 9 (Fig. 8.4).

8.10 Classification of Accommodation Units

Classification of accommodation units in various tourism places as star, non-star, and others are described in Table 8.3. Total star hotels are only 25 out of which the highest number of star hotels is located in Mussoorie and Dehradun towns/cities, which is more than six. CNP has a total of six hotels with stars. In all other selected places, star hotels are less than three. The total number of non-star hotels is 548 out of which, Mussoorie and Nainital have more than 100, followed by Dehradun. The

Table 8.3 Classification of accommodation units as star, non-star, and others

Indices	Levels	Towns/cities
Star hotels (25)		
> 6	High	Mussoorie and Dehradun
3–6	Medium	CNP
<3	Low	Ranikhet, Kausani, Bageshwar, Nainital, Uttarkashi, Pauri, Almora, Pithoragarh, Kotdwar, Joshimath, Yamunotri, Gangotri, Badrinath, Rishikesh, and Haridwar
Non-star hotels (548)		
<100	High	Mussoorie and Nainital
50–100	Medium	Dehradun
>50	Low	Haridwar, Joshimath, Rishikesh, Uttarkashi, Pauri, Kausani, Gangotri, Pithoragarh, Kotdwar, Yamunotri, Badrinath, Almora, Bageshwar, Ranikhet, and CNP
Others (1195)		
<100	High	Haridwar
50–100	Medium	Yamunotri, Uttarkashi, and Gangotri
>50	Low	Almora, Rishikesh, Mussoorie, Dehradun, Badrinath, Nainital, Joshimath, Ranikhet, Kotdwar, Pithoragarh, Pauri, Bageshwar, CNP, and Kausani

other natural and cultural places have less than 50 non-star hotels. Other types of accommodation are 1195 in number. Of which, Haridwar has the highest number, which is categorized as high, followed by Yamunotri, Uttarkashi, and Gangotri with 50–100 numbers. Less than 50 accommodations are available in other tourists' places (14 towns/cities) (Fig. 8.5).

Fig. 8.5 Accommodation— star and non-star hotels

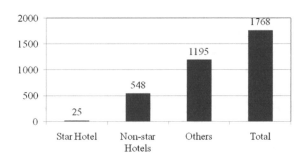

Uttarakhand Tourism Development Board	
Garhwal Mandal Vikas Nigam	Kumaon Mandal Vikas Nigam
Seven District Tourism Department	Six District Tourism Department
District Tourism Development Officer	District Tourism Development Officer
Total 38 Guest Houses in Tourist Places of Garhwal Region	Total 35 Guest Houses in Tourist Places of Kumaon Region
Each Guest House has a Tourist Officer and other ministerial staffs	Each Guest House has a Tourist Officer and other ministerial staffs

Fig. 8.6 Structure of tourism institutions in the Uttarakhand Himalaya

8.11 Institutional Facilities

Until 2000, the state of Uttarakhand was a division of Uttar Pradesh and it was known as the UP Hills. During the time, there was a Tourism Department located at Dehradun, which was planning and coordinating the tourism development of UP Hills. The resources were limited and tourism development could not take shape although, tourism has been a centuries-old activity, mainly pilgrimage tourism. In 2000, Uttarakhand got statehood, carving out from Uttar Pradesh. Sooner, in 2001, UTDB was established. The major objective of UTDB was to enhance the institutional capacity for tourism development. Already, the GMVN with its headquarter at Dehradun and KMVN with its headquartered at Nainital was functional. However, these developmental agencies have geared up and several other policy measures were taken for tourism development in Uttarakhand Himalaya after its formation. One of the measures was the introduction of homestay tourism both in urban and rural areas. At present, the structure of tourism institutions is as follows (Fig. 8.6).

UTDB has two developmental bodies—GMVN and KMVN. Both have tourism departments at the district level, seven in Garhwal region, and six in the Kumaon region. Further, GMVN has 38 guest houses in tourist places and pilgrim centers of the Garhwal region and KMVN has 35 guest houses in tourist places and pilgrim centers of the Kumaon region. Each guest house has a tourist officer and other ministerial supporting staff.

8.12 Medical Facilities

The Uttarakhand Himalaya has a feasible healthy climate mainly in the summer season. In the meantime, the climate varies from the river valleys to the middle-altitudes, and the highlands. Because the entire region is mountainous and the terrain is precipitous and rugged, medical facilities are a priority area. However, on the road routes, medical facilities are not adequate. The service centers, where medical

Table 8.4 Hospitals in Uttarakhand (Dec 2018)

Types of facility	Number
Sub-centres	1918
Primary health centers	301
Community health centers	73
Sub-district hospitals	47
District hospitals	21
Total	2360

facilities are available, are sparsely located. Often, tourists/pilgrims face acute health problems. Table 8.4 shows hospital facilities in Uttarakhand in 2018. A total of 2360 hospitals are available in Uttarakhand. The highest number of 1918 is available in sub-centers which is 81.27%. Other health facilities are primary health centers, community health centers, sub-district hospitals, and district hospitals, which are 18.73% of the total.

Source Directorate of Economics and Statistics, Government of Uttarakhand, Dehradun

8.13   Educational Facilities

Education is an important element of tourism development. Tourism needs human resource—skilled workers, educated tour guides and provision of other services. Therefore, a good skill based education system is required in the State. Uttarakhand obtains a good education level however the educated youth are migrating out of state. As a result, skilled labour availability is insufficient. Uttarakhand has many national-level educational institutions as Universities, IIT and Medical Colleges. Under the universities, academic tourism departments are established. In all over India and worldwide, the trained and skilled workers from Uttarakhand are working mainly in hotel industry. Uttarakhand cuisine is very nutritional and tasty. However, skilled hotel workers are out-migrated from the state. Here, trained tourism guides are scarcely available.

8.14   Conclusions

In this chapter, infrastructural facilities of Uttarakhand tourist and pilgrim centers— transportation, accommodation and institutional are elaborated extensively. It has been noticed that although roads are traversed in the valleys and middle altitudes yet the highlands areas are still inaccessible. Many places of tourists/pilgrims' interests are not well connected by any means of transportation. Further, the condition of roads is not satisfactory to support tourism throughout the year. Many roads are traversed along with the fragile landscapes, where landslides are very common. This leads to roadblock and accidents. The railway services are available only in the plain regions. Further, the construction of the rail route along the fragile slope is not advisable. Air transport is suitable; however, there are only five airports and a helipad, of which only Dehradun airport is connected with the major cities of India. Construction of roads is possible through pillars and tunnels, which will minimize roadblocks and accidents. The government has already initiated the construction of railways in the four highland pilgrimages, which will enhance tourism development. The major tourist places can be connected by the airways and the fare can be subsidized by the government to promote air transportation.

Accommodation facilities are not adequate. However, the tourists/pilgrims' inflow is very high. As a result, the tourists/pilgrims plan their return from the tourist places/pilgrimages for the same day. Homestay facilities can be developed in the rural and urban areas, where the major tourist places and pilgrimages are located. The income level of tourists/pilgrims varies. Many of them have a low-income level therefore, a range of accommodation can be provided keeping the income level of the pilgrims and tourists in mind. Institutional facilities need to be enhanced. There are several places of tourist interest, which are not yet explored. UTDB can play a vital role in developing lesser-known places with the help of GMVN and KMVN. If all the infrastructural facilities are developed and provided to tourists/pilgrims, then the Uttarakhand Himalaya may be a hub of tourism. Subsequently, it will lead to economic development through income generation and employment augmentation.

References

Adobayo KA, Iweka COA (2014) Optimizing the Sustainability of tourism infrastructure in nigeria through design for deconstruction framework. Am J Tourism Manage 3(1A):13–19

Cooper C, Fletcher J, Fyall A, Gilbert D, Wanhill S (2008) Tourism principles and practice. Pearson Education Limited, England

Crouch G, Ritchie JRB (1999) Tourism, competitiveness, and societal prosperity. J Bus Res 44(3):137–152. https://doi.org/10.1016/S0148-2963(97)00196-3

Crouch G, Ritchie JRB (2000) The competitive destination: a sustainability perspective. Tourism Manage 21(1):1–7

Dwyer L, Forsyth P, Dwyer W (2010) Tourism economics and policy. Channel View Publications, Bristol

Formica S, Uysal M (1996) Market segmentation of an international cultural-historical event in Italy. J Travel Res 36(4):16–24. https://doi.org/10.1177/004728759803600402

Fourie J, Santana GM (2011) The impact of mega-sport events on tourist arrivals, Stellenbosch University, Department of Economics. Viewed, Mar 2018. https://econpapers.repec.org/paper/szawpaper/wpapers119.htm

Garay L, Canoves G (2011) Life cycles, stages and tourism history. Ann Tourism Res **38**(2). https://doi.org/10.1016/j.annals.2010.12.006

Getz D (1992) Tourism planning and destination life cycle. Ann Tourism Res 19(4):752–770. https://doi.org/10.1016/0160-7383(92)90065-W

Hadzik A, Grabara M (2014) Investments in recreational and sports infrastructure as a basis for the development of sports tourism on the example of spa municipalities. Pol J Sport Tourism 21:97–101. https://doi.org/10.2478/pjst-2014-0010

Hansen NM (1965) Unbalanced growth and regional development. Western Econ J 4:3–14

Heath RA (1992) Wildlife-based tourism in Zimbabwe: an outline of its development and future policy options. Geogr J Zimbabwe 23:59–78

Khadaroo J, Seetanah B (2008) Transport infrastructure and tourism development: a dynamic gravity model approach. Tourism Manage 29:831–840. In: Jafari J, Xiao H (eds) Encyclopedia of tourism. Springer Reference, Switzerland

Lohmann G, Netto AP (2017) Tourism theory concepts, models and systems. CABI, Oxfordshire

Lovelock J, Lovelock K (2013) The ethics of tourism: critical and applied perspectives. Routledge, New York

McConnell KH (1985) The economics of outdoor recreation. In: Kneese AV, Sweeney JL (eds) Handbook of natural resources and energy economics, vol 11. Elsevier Science Publisher, New York, pp 678–722

Mera, K. (1973) Regional production functions and social overhead capital: An analysis of the Japanese case, Regional and Urban Economics, Vol. 3, pp. 157–186. https://doi.org/10.1016/003 43331(73)90013-4

Murphy P, Pritchard MP, Smith B (2000) The destination product and its impact on traveler perceptions. Tourism Manage 21(1):43–52

Nest & Wings (2010) Uttarakhand hotels guide. Nest & Wings, New Delhi

Popesku J (2011) Menadžment turističke destinacije. Univerzitet Singidunum, Beograd

Ritchie JRB, Crouch GI (2005) The competitive destination: a sustainable tourism perspective. CABI, Wallingford

Sati VP (2018) Carrying capacity analysis and destination development: a case study of Gangotri Tourists/Pilgrims' circuit in the Himalaya. Asia Pacific J Tourism Res 23(3):312–322. https://doi.org/10.1080/10941665.2018.1433220

Sakai M (2006) Public sector investment in tourism infrastructure. In: Dwyer L, Forsyth P (eds) International handbook on the economics of tourism. Edward Elgar, Cheltenham, pp 266–279

Sharpley R (2009) Tourism development and the environment: beyond Sustainability? Earthscan, New York

Swarbrooke J, Horner S (2001) Business travel and tourism. Butterworth-Heinemann, Jordan Hill

Tourism and Transport Forum (TTF) (2012) Tourism infrastructure policy and priorities. https://www.ttf.org.au/Content/infprio201112.aspx. Accessed 20 Mar 2015

UNWTO (2007) A practical guide to tourism destination management. https://www.e-unwto.org/doi/book/10.18111/9789284412433

Chapter 9
A Sustainable Homestay Tourism and Its Prospects

9.1 Introduction

The homestay tourism is a kind of tourism, which provides an affordable and comfortable stay to the tourists/pilgrims in the major tourists/pilgrims destinations. It involves the local community in providing services and thus it enhances livelihoods. In homestay tourism, local communities/families host tourists/pilgrims, offer a pleasant stay (Frederick 2003), and in return, they earn money, which sustains their income and livelihood. Homestay tourism also helps in strengthening the local culture and customs. Food and beverages in a homestay are usually included and duration-may be daily, weekly, or monthly (Rivers 1998). The homestay tourism is very beneficial in terms of preserving the local authentic heritage, employment augmentation and income generation (Wang 2007).

Homestay tourism is managed by individuals or communities (Timilsina 2012) and it offers tourists the new experience of hospitality and enhances the income and economy of the rural people (Gangotia 2013). The local people can conserve the environment and their culture through homestay tourism (Laurie et al. 2005). The homestay tourism can be the major means of poverty reduction and rural livelihood enhancement as it integrates all tourism activities such as trekking, cultural tourism, agro-tourism, health tourism and eco-tourism (Devkota 2010). Besides, it provides an opportunity to sell the local agricultural and milk products (Budhathoki 2013). A tourist learns about indigenous knowledge and traditional cultures, interacts with the local community, and spends a comfortable/homely stay during homestay tourism (Wijesundara and Gnanapala 2015). On the other hand, the homestay tourism generates income and employment to the host and enhances the rural economy and raise awareness (Lynch et al. 2009; Jamilah and Amran 2007).

The homestay tourism has multiplier impacts on economy and development, occupation, lifestyle, and population structure (Beeton 2006; Gangte 2011; Guo and Huang 2011; Kerstetter and Bricker 2009; Pizam and Milman 1986). It has

V. P. Sati, *Sustainable Tourism Development in the Himalaya: Constraints and Prospects*, Environmental Science and Engineering, https://doi.org/10.1007/978-3-030-58854-0_9

economic, social and cultural benefits; it improves the quality of life of the host community (Bhuyian et al. 2011). It can also empower communities through development interventions (Singh et al. 2003) as it requires low capital outlay thus, it can be accessed by the poor (Rea 2000).

The homestay tourism was initiated first in Malaysia in 1998 with its aim to provide local hospitality to tourists and to enhance the income and livelihood of the host (Lowry 2016). The other countries of Southeast Asia such as Vietnam, Malaysia, Thailand, and Indonesia promote successful homestay tourism (Ghimire 1995 and Kwaramba et al. 2012). In the Mekong region of Asia, homestay tourism has helped communities to reduce the incidence of rural poverty (Leksakundilok 2004). The cultural aspect of homestay tourism is a key component. Further, the interaction between host and guest is the core product, while education, entertainment, and enrichment activities are important contributors (Kayat 2010).

Tourism contributes about 50% of the total GSDPs in the Uttarakhand Himalaya (UTDB 2018). It supports livelihood to a large population of the state and plays an important role in the state's socio-economic goals as the state moves on its vision to become a green economy. Hotel industries are developed in the major tourists' places and pilgrimages, which are owned by the businessmen. The rural youth are serving in these hotels as cooks and cleaners. However, hotel facilities in the small and remote places are lagging and rural people do not receive any benefits of tourism and they remain poor (Sati 2013). The tourists/pilgrims both domestic and foreigners are also deprived of exploring such beautiful places. These drivers manifested the state government to draft 'Deen Dayal Upadhyaya Griha Awaas Home Stay (DDUGAHS) regulation 2018'. The main purpose of the regulation 2018 was to provide an affordable homestay to domestic and foreign tourists through the involvement of the local community or an individual basis. It has further aimed to enhance accommodation facilities, augmentation of employment, and additional sources of income to a community or an individual at the village level.

Uttarakhand has a wide range of rural communities with varying cultures. Dwellings/huts of these communities in the highlands and alpine pasturelands provide uniqueness to the homestay for tourists/pilgrims. Further, the diversity in cultural and natural aspects in this region also manifests its suitability for homestay tourism. The idea of homestay tourism is based on walking and slow tourism developed by the state government. It gives the experience of the different culture, lifestyle and food cuisines of the distant villages. The nature of homestay tourism is more comfortable, exciting, engaging and educative. It is tourism for peace from the crowded hotels. It is comparatively affordable, less expensive than a hotel stay.

Uttarakhand has numerous natural and cultural locales. It has-a high potential for developing homestay tourism. Although tourism has a greater role in the state's economy, yet its benefits did not reach the poor rural people who are the main tourism service providers. The homestay tourism is the new and emerging area for researching the Uttarakhand Himalaya. Here, no one has carried out any substantial work on homestay tourism so far. This chapter examines the sustainable homestay tourism in the Uttarakhand Himalaya. It analyses the prospects of homestay tourism in terms of income generation, employment augmentation and livelihood enhancement in rural areas. This study proposes sustainable homestay tourism and describes its impact on culture, economy, and livelihood.

9.2 Homestay Tourism Policy in Uttarakhand

The homestay tourism policy in Uttarakhand was started in 2015. However, it got impressive progress in 2018 when it was renamed as 'DDUGAHS Regulation 2018'. This regulation was framed and implemented under the aegis of the UTDB, which was established by the state government in 2001. The scope of the policy is to implement it in the entire state and in the areas, which do not come under Municipal Corporation. Homestay is inspired by the residential unit, which is completely a 'residential premise' in which the house owner also resides with his family. The Chief Executive Officer of the UTDB is the head of DDYGAHS. A house/landowner, the permanent resident of Uttarakhand, who has enough space for homestay (one to six rooms) in his residential premises, can apply for the DDYGAHS with condition that he will supply food and beverages to the tourists/pilgrims. The owner must have registered his house for homestay and he/she should not be a bank defaulter. Under this policy, the state government will exempt the owners from paying taxes up to three months and will provide free electricity and water for this period. The owner can change the land use from residential to non residential without any legal formalities. Initially, a lump sum amount (capital subsidy) will be given to the owner for the construction of rooms and repairing them. There is a selection committee, headed by the District Magistrate along with five members including District Tourism Development Officer for the selection of beneficiaries. The state's reservation policy is also be adopted in this scheme. If there are six or more than six homestays in a village, they are called 'Homestay Cluster'. Infrastructural facilities such as the construction of website, internet, online-offline marketing facilities, publicity, rating, training and the loan will be provided to the owners by UTDB. All activities will be monitored by UTDB at the district level. Recently, the state government has converted abandoned school buildings for homestay to boost tourism mainly those are located on the way to the highland pilgrimages and along the trekking routes. Under the vision 2020, the policy has set a target to develop around 5000 homestay facilities.

Both qualitative and quantitative methods were used to present data in this chapter. First, the data were gathered from secondary sources mainly from UTDB, Dehradun, on homestay facilities—registered units, total rooms, and total beds in urban and rural areas at the district level from 2017 to 2019. The income from the homestay tourism was calculated at per unit accommodation and finally, the income from homestay tourism was calculated at the district level. Change in homestay facilities from 2017 to 2019 was also obtained. All these data—homestay facilities, income generated from it and change were analyzed and presented through graphs. Descriptive statistics were used to analyze homestay registered units and income obtained from it, from 2017 to 2019, in both urban and rural areas. A sustainable homestay tourism model was made to describe the cultural, natural, and human resource aspects of homestay tourism development. The impact of homestay tourism on enhancing rural livelihoods, generating income, augmenting employment and cultural strengthening was illustrated. The author has rapidly visited the major tourists/pilgrims' destinations and observed that homestay tourism has very high potential in enhancing rural economy and livelihood in the Uttarakhand Himalaya.

9.3 District-Wise Registered Units and Income from Homestay Tourism in Urban Areas

District-wise units and income from homestay tourism in urban areas in the years 2017, 2018, and 2019 were analyzed (Table 9.1). Each registered unit has an average of five double bed rooms. In the urban areas, per night/room cost is average 1200 Indian Rupees (INR) with breakfast. In 2017, the highest registered units were obtained by Dehradun district (49), followed by Nainital (21). Haridwar (4), Almora (3), USN (2), and Champawat (1) also has homestay tourism in the urban areas. The other districts—Bageshwar, Chamoli, Pauri, Pithoragarh, Rudraprayag, and Uttarkashi did not have any homestay facilities. Likely, the income from the homestay tourism was the highest in Dehradun district with INR 294 thousand, followed by the Nainital district (126 thousand). The other four districts have less than INR 30

Table 9.1 District-wise registered units and income from homestay tourism in urban areas

District	2017		2018		2019	
	Registered units	Income (INR)	Registered units	Income (INR)	Registered units	Income (INR)
Almora	3	18,000	3	18,000	3	18,000
Bageshwar	Nil	Nil	Nil	Nil	1	6000
Chamoli	Nil	Nil	Nil	Nil	15	90,000
Champawat	1	6000	2	12,000	2	12,000
Dehradun	49	294,000	64	384,000	182	1,092,000
Haridwar	4	24,000	4	24,000	12	72,000
Nainital	21	126,000	25	150,000	39	234,000
Pauri	Nil	Nil	Nil	Nil	Nil	Nil
Pithoragarh	Nil	Nil	Nil	Nil	Nil	Nil
Rudraprayag	Nil	Nil	Nil	Nil	3	18,000
Tehri	Nil	Nil	2	12,000	2	12,000
USN	2	12,000	2	12,000	2	12,000
Uttarkashi	Nil	Nil	V	Nil	6	36,000
Total	80	480,000	102	612,000	267	1,602,000

Source UTDB (2019)

Fig. 9.1 Homestay facilities in urban areas 2017–2019

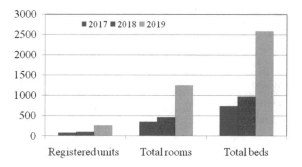

thousand income from homestay tourism. In 2018, the registered units in Dehradun district increased to 64 with INR 384 thousand income. In the Nainital district, the units increased to 25, and income was INR 150 thousand. Champawat received one additional unit and income was just double whereas, in Almora, Haridwar and USN districts, the figure was unchanged. Meanwhile, the Tehri district started homestay tourism with two registered units. The registered units and income were three times higher in Dehradun district (182 units and 10.92 million INR respectively) in 2019. However, the Nainital district got a marginal increase in registered units (39) and income (INR 234 thousand). Although the registered units increased three times (from 4 to 12) in the Haridwar district yet the total number of units was less. Chamoli (15), Uttarkashi (6), and Rudraprayag districts registered homestay units the first time and accordingly earned income.

Total homestay facilities in Uttarakhand as registered units, total rooms, and beds in urban areas 2017 to 2019 is shown in Fig. 9.1. In comparison to 2017, homestay facilities increased significantly in 2018 and then in 2019. The registered units were 80 in 2017, which increased to 102 in 2018 and 267 in 2019. Subsequently, total rooms and beds increased during the period.

9.4 District-Wise Registered Units and Income from Homestay Tourism in Rural Areas

Rural areas have more homestay facilities for tourists/pilgrims than in urban areas. At the district level, Tehri district had the highest 65 registered units and INR 325 thousand income in 2017, followed by district Almora with 62 registered units and INR 310 thousand income. Uttarkashi and Nainital districts had 19 registered units each with INR 95 thousand income (Table 9.2). Bageshwar and Dehradun districts had 14 and 13 registered units, respectively. The other districts Pithoragarh, Chamoli, Pauri, Champawat, Haridwar, and Rudraprayag had less than 10 registered units and they earned income subsequently. The USN district did not have any registered unit in homestay tourism. There was a slight increase in registered units and income in a few districts in 2018. Tehri district retained the first place with 77 registered units

Table 9.2 District-wise registered units and income from homestay tourism in rural areas

District	2017		2018		2019	
	Registered units	Income (INR)	Registered units	Income (INR)	Registered units	Income (INR)
Almora	62	310,000	66	330,000	100	500,000
Bageshwar	14	70,000	14	70,000	28	140,000
Chamoli	2	10,000	2	10,000	111	555,000
Champawat	1	5000	1	5000	3	15,000
Dehradun	13	65,000	15	75,000	29	145,000
Haridwar	1	5000	1	5000	1	5000
Nainital	19	95,000	27	135,000	110	550,000
Pauri	2	10,000	4	20,000	21	105,000
Pithoragarh	6	30,000	6	30,000	141	705,000
Rudraprayag	1	5000	1	5000	54	270,000
Tehri	65	325,000	77	385,000	103	515,000
USN	Nil	Nil	Nil	Nil	Nil	Nil
Uttarkashi	19	95,000	19	95,000	54	270,000
Total	205	1,025,000	233	1,165,000	755	3,775,000

Source UTDB (2019)

and INR 385 thousand income followed by Almora district with 66 units and INR 330 thousand income. Registered units increased to 27 in the Nainital district and 15 in Dehradun district. The other districts have almost no change in registered units and subsequently in income. In the meantime, a tremendous increase was noticed in registered units and income in homestay tourism in 2019. Pithoragarh district jumped and got first place with 141 units and INR 705 thousand income, followed by Chamoli (111 units) and Nainital (110 units) districts. Tehri was in fourth place with 103 units and INR 515 thousand income. The Almora district registered 100 units and earned INR 500 thousand as income. The two districts—Rudraprayag and Uttarkashi registered 54 units each and earned subsequent income. Haridwar and Champawat districts have the lowest registered units (1 and 3 units, respectively).

Figure 9.2 shows total homestay facilities in Uttarakhand—registered units, total rooms, and beds in rural areas from 2017 to 2019. In comparison to urban areas, homestay facilities in rural areas are about three times higher during the period and it has increasing trends. A total of 205 registered units were available in 2017, which has increased to 233 registered units in 2018. In 2019, the total of homestay registered units was 755. Subsequently, the total rooms and beds increased.

Fig. 9.2 Homestay facilities in rural areas 2017–2019

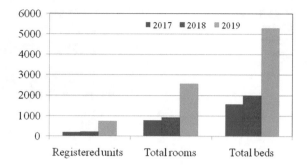

9.5 Statistical Analysis of Homestay Tourism in Urban and Rural Areas

Descriptive statistics—minimum, maximum, and mean value and std. deviation of registered homestay units and incomes earned from it from 2017 to 2019 both in urban and rural areas, were analyzed (Table 9.3). It has been noticed that the homestay registered units were less in urban areas than in rural area and subsequently income

Table 9.3 Statistical analysis of homestay tourism in urban and rural areas

Variables	Minimum	Maximum	Mean	Std. Deviation
Urban areas				
Registered units (No.)				
2017 (n = 6)	1	49	13.33	19
2018 (n = 7)	2	64	14.57	23.35
2019 (n = 11)	1	182	24.27	53.48
Income (INR)				
2017 (6)	6000	294,000	80,000	114,011
2018 (7)	12,000	384,000	87,429	140,108
2019 (11)	6000	1,092,000	145,636	320,875
Rural areas				
Registered units (No.)				
2017 (n = 12)	1	65	17.08	22.77
2018 (n = 12)	1	77	19.42	25.82
2019 (n = 12)	1	141	62.92	47.97
Income (INR)				
2017 (12)	5000	325,000	85,417	113,827
2018 (12)	5000	385,000	97,083	129,078
2019 (12)	5000	705,000	314,583	239,872

Source By the author

Fig. 9.3 Increasing
homestay accommodation
facilities in urban and rural
areas

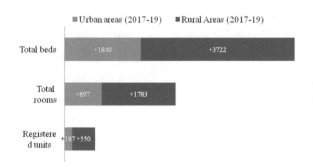

from the homestay tourism varies in both areas. However, the trend of the homestay
registered units and income is increasing.

9.6   Increasing Homestay Facilities

After revised homestay tourism regulation 2018, there was a substantial increase in
homestay accommodation both in urban and rural areas in Uttarakhand. Figure 3
shows an increase in homestay facilities from 2017 to 2019. In urban areas, an
increase of 187 was noticed in registered units. A total of 897 rooms and 1840 beds
were increased during the period. In rural areas, an increase in homestay facilities
was higher than in urba areas. A total of 550 registered units were increased during
the period. A total of 1783 rooms were increased and a total of 3722 beds were also
increased (Fig. 9.3).

9.7   Income Obtained from Homestay Tourism

Income obtained from homestay tourism was analyzed both in urban and rural areas.
In 2019, in urban areas, a total of 267 households were involved in-homestay tourism
whereas it was 755 households-involved in homestay tourism in rural areas. In 2017,
the figures were 80 households in urban areas and 205 households in rural areas. The
rate of homestay-including breakfast was average 1200 INR/night/roomin urban
areas whereas it was INR 1000/night/room in rural areas. In 2017, the total income
from homestay tourism in urban areas was less than INR 500 thousand; it increased
to INR 0.6 million in 2018 and 1.6 million in 2019. In terms of income from the
homestay tourism in rural areas, it was 1 million in 2017, about 1.2 million in 2018,
and 3.8 million INR in 2019. The increase in income from the homestay tourism
between 2017 and 2019 was three times higher in urban areas whereas it was about
four times higher in rural areas (Fig. 9.4).

Fig. 9.4 Income obtained from homestay tourism (INR)

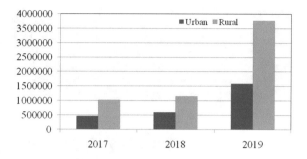

9.7.1 A Sustainable Model of Homestay Tourism

Uttarakhand practices diverse tourism—natural, cultural, adventurer and wildlife/park/ecotourism. The tourists/pilgrims both domestic and international visit these places every year. Infrastructural facilities—accommodation, transportation, and institutional—are inadequate. Accommodation facilities are mostly owned by businessmen from and outside of Uttarakhand. The local people are not much involved in providing accommodation services and are not earning money out of it. From economic point of view, the Uttarakhand Himalaya is underdeveloped. Agriculture is the main occupation of the people with low output and therefore an exodus number of people have out-migrated. On the other hand, it has a rich cultural heritage. The region is rich in organizing fairs and festivals, having diverse-food and beverages, and making arts and crafts. The highland and river valleys pilgrim centers are the major attractions of the pilgrims. Similarly, natural locales are plenty. However, the development of the sacred places and benefits of tourism did not reach to the poor rural people. Development of homestay tourism will boost rural livelihoods, sustain the economy and minimize migration, conserve local culture and tradition. Figure 9.5 presents a sustainable homestay tourism model, which includes cultural aspects, physical aspects, and human resource aspects of homestay tourism. The homestay tourism can be sustainable with performing culture and customs, celebrating fairs and festivals, and to publicizing and strengthening natural and cultural places. Adequate infrastructural facilities like distinctive accommodation including modern toilets, home services, safety, local knowledge, home-cooked food, unique activities, celebrating festivals, internet, owner's website, and mobile apps can be ensured to attract different income level tourists/pilgrims.

9.8 The Impact of Homestay Tourism

The most important impact of homestay tourism is the possibility of a big boost to rural livelihoods (Fig. 9.6). Since homestay tourism is practiced in the rural areas and by the local people; the benefits from the homestay tourism will ultimately enhance rural economy and livelihood. In Uttarakhand, rural out-migration has

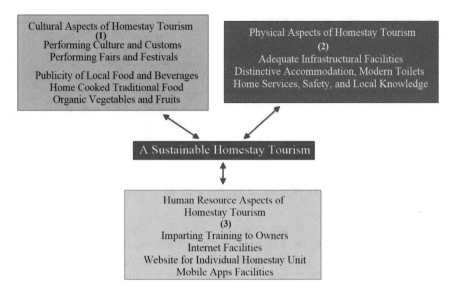

Fig. 9.5 Sustainable homestay tourism

Fig. 9.6 Impact of homestay tourism

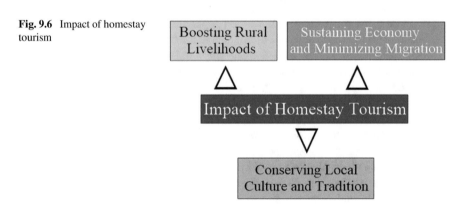

become a major impediment, which has led to depopulation in rural areas. Economically, rural areas are underdeveloped. The development of homestay tourism will increase income and economy, augment employment, and minimize out-migration. Conserving local culture and tradition is another important impact of homestay tourism. The Uttarakhand Himalaya has rich culture and customs. It can be conserved through homestay tourism. The local people can serve traditional food and beverages to tourists/pilgrims during their stay. They can perform arts and culture.

9.9 Discussion

Nestled in the lap of the Himalaya, the Uttarakhand Himalaya has spectacular land-scapes—the river valleys, middle altitudes, highlands, the alpine pasturelands, and the snow-clad mountain peaks, which has tremendous scope of tourism development. The highland and river valley pilgrimages are world-famous and well developed where hundreds of thousands of pilgrims visit every year (Sati 2015). In the meantime, the region has an agricultural economy. About 70% of the population is living in rural areas and practicing subsistence agriculture. The output from traditionally grown crops is not enough to feed the people's daily needs sustainably. In the remote and inaccessible areas, food security and malnutrition are common. A big segment of rural people (about 22%) is living below the poverty line (Statistical Diary 2018). Because of all these impediments, an exodus number of people out-migrated to the urban areas within or outside the state for the search of jobs and better livelihoods. The development of homestay tourism will lead to income generation, employment augmentation and enhancement of rural livelihoods. It will also minimize out-migration from rural areas.

Tourists/pilgrims inflow in the Uttarakhand Himalaya is very high. In 2018, the total pilgrims visited in the pilgrimages of Uttarakhand were 25.26 million. Similarly, the total tourists visited in the natural locales were 4.48 million (UTDB 2019), as a result, tourism shares 50% of GSDP of the state. The state has rich cultural heritage. Numerous fairs are organized and festivals are celebrated every year. The homestay tourism can strengthen rich culture as the tourists/pilgrims can participate in celebrating festivals and be part of fairs during their homestay.

In Uttarakhand, a few homestay facilitiesare available mostly in the major trekking routes (Macek 2012). Although the homestay tourism regulation was implemented in 2015, it got momentum in 2018 when-DDUGAHS Regulation 2018 was implemented. Data on homestay facilities—registered units, rooms, and beds from 2017 to 2019 shows that there was not much increase in a homestay in 2017–2018 whereas, in 2018–19, the increase in homestay facilities was two to three times higher. It has also been noticed that the homestay facilities increased more in rural areas than in urban centers during the period. The reason is that urban areas have enough accommodation of various types. In rural areas, accommodation facilities for tourists/pilgrims are limited and thus homestay may be a suitable alternate accommodation. A study depicts that about 8349 pilgrims visit Gangotri temple per day during the peak season; however, accommodation facility is available for only 3222 persons, and therefore, about 59% pilgrims return to Uttarkashi on the same day (Sati 2018).

At-the district level, the homestay tourism in urban areas is practiced in the districts, which are mostly urban and located in plain areas. The urban population is higher in these districts such as Dehradun and Nainital. On the other hand, the homestay tourism in the rural areas is available mostly in mountainous districts, where many tourists' places/pilgrimages are immensely located. Tehri and Almora districts have the highest homestay in rural areas. However, the homestay in both urban and rural areas is just negligible to accommodate a huge number of tourists/pilgrims.

Homestay tourism has high prospects mainly in rural areas of the Uttarakhand Himalaya. Several tourists' places and pilgrimages are inaccessible, can be approached through trekking only. The homestay will be the most suitable form of accommodation in these trekking routes. In Homestay tourism, an owner can facilitate one to six suits for the guests. Any village, which will have more than six homestay facilities, can be developed as a cluster and the infrastructural facilities—internet, owner website, and mobile apps—can be provided by the state government. Training can be imparted to the homestay owner to earn knowledge on keeping high standards of the hospitality industry—to treat visitors with courtesy, maintain hygiene, and have their residential premises equipped with necessary amenities—internet and roads. As per the policy, the homestay will be developed in the villages, which are along the main trekking routes. Uttarakhand has several trekking routes to reach tourists' places and pilgrimages where many domestic and foreign tourists/pilgrims visit every year. Proper sites can be identified to develop homestay along the trekking routes.

9.10   Conclusions

Homestay tourism is an emerging sector for economic growth and rural livelihood enhancement in the Uttarakhand Himalaya as it has significant prospects. The places of tourists/pilgrims' interest can be developed and the rich culture and delightful delicious cuisine of the state can be popularized through homestay tourism. The homestay tourism can also augment employment, make the rural people self-reliance, and can check/minimize the forced rural-urban migration. The homestay policy will be a tool for developing Uttarakhand as a priority tourist location. Homestay may not be successful without government support, because it needs lots of infrastructural facilities and regulation. Therefore, the government should play a greater and rigourous role to develop homestay tourism in the state. In Uttarakhand, the hotel industries are developed in major cities, tourist places, and pilgrimages and owned by the businessmen of within and outside of the state.Rural areas and the poor local people, who are tourism service providers are not reaping/enjoying the benefits of tourism development in the state. The development of homestay tourism will manifest strengthening rural economy and livelihood. The state government should ensure that the benefits of homestay tourism reach to the local people in improving their livelihood and living standards.

References

Beeton S (2006) Community development through tourism. Land Links, Collingwood
Bhuyian MH, Ismail SM, Islam R (2011) Potentials of Islamic tourism: a case study of Malaysia on East Coast Economic Region

Budhathoki B (2013) Impact of homestay tourism on livelihood: a case study of Ghale Gaun, Lamjung, Nepal. A thesis submitted to Norwegian University of Life Sciences

Devkota T (2010) Gorkhaparta—the rising Nepal. Retrieved on 20 Feb 2011 from http://www.gor khapatra.org.np/rising.detail.php?article_id=45767&cat_id=7

Frederick CM (2003) Merriam-Webster's collegiate dictionary. Merriam-Webster, p 595

Gangotia A (2013) Home stay scheme in Himachal Pradesh: a successful story of community based tourism initiatives (CBTIS). Global Res Anal 2(2):206–207

Gangte M (2011) Sustainable community development alternative: unlocking the lock. J Sustain Develop 4(2):61–71

Ghimire G (1995) Case study on the effects of tourism on culture and the environment in Nepal, UNESCO. Principal Regional Office for Asia and the Pacific, Bangkok

Guo W, Huang Z-F (2011) Study on the development of community power and functions under the background of the development of rural tourism. Tourism Tribune/Luyou Xuekan 26(12):83–92

Jamilah J, Amran H (2007) KPW and women roles in Banghuris Homestay. Rural Tourism Research, Malaysia

Kayat (2010) The Nature of cultural contribution of a community-based homestay programme 2010

Kerstetter D, Bricker K (2009) Exploring Fijian's sense of place after exposure to tourism development. J Sustain Tourism 17(6):691–708

Kwaramba HM, Lovett JC, Louw L, Chipumuro J (2012) Emotional confidence levels and success of tourism development for poverty reduction: The South African Kwam Makana home-stay project. Tourism Manage 33(4):885–894

Laurie ND, Andolina R, Radcliffe SA (2005) Ethnodevelopment: social movements, creating experts, and professionalizing indigenous knowledge in Ecuador. In: Laurie ND, Bondi L (eds) Working the spaces of neoliberalism: activism, professionalization, and incorporation

Leksakundilok A (2004) Ecotourism and community-based ecotourism in the Mekong Region. The University of Sydney. Australian Mekong Resource Centre, Sydney

Lowry L (2016) Business & economics. The Sage International Encyclopedia of Travel and Tourism

Lynch PA, McIntoch AJ, Tucker H (2009) Commercial homes in tourism: an international perspective. Routledge, London

Macek IC (2012) Homestays as livelihood strategies in rural economies: the case of Johar Valley, Uttarakhand, India. A thesis submitted to the University of Washington, Washington

Pizam A, Milman A (1986) The social impacts of tourism. Tourism Recreat Res 11:29–33

Rea MH (2000) A Furusa to away from home. Ann Tourism Res 27(3):638–660

Rivers WP (1998) Is being their enough? The effects of homestay placement on language gain during study abroad. Foreign Lang Ann 31(4):492–500

Sati VP (2018) Carrying Capacity analysis and destination development: a case study of Gangotri Tourists/Pilgrims' circuit in the Himalaya. Asia Pacific J Tourism Res 23(3):312–322. https://doi.org/10.1080/10941665.2018.1433220

Sati VP (2015) Pilgrimage tourism in mountain regions: socio-economic implications in the Garhwal Himalaya. South Asian J Tourism Heritage 8(1):164–182

Sati VP (2013) Tourism practices and approaches for its development in the Uttarakhand Himalaya, India. J Tourism Challenges Trends 6(1):97–112

Singh S, Timothy DJ, Dowling RK (2003) Tourism and destination communities. CABI Publishing, Oxon

Statistical Diary (2018) Uttarakhand statistical diary, Dehradun

Timilsina P (2012) Homestay tourism boosts GhaleGaon's economy. Retrieved from http://www.gorkhapatra.org.np./rising.detail.php?Article_id=23200&cat_id=4. On 10 June 2013 at 11:52.

UTDC (2019) Annual report. Uttarakhand Tourism Development Council, Dehradun, Uttarakhand

UTDC (2018) Annual report. Uttarakhand Tourism Development Council, Dehradun, Uttarakhand

Wang Y (2007) Customized authenticity begins at home. Ann Tourism Res 34(3):789–804

Wijesundara N, Gnanapala Athula C (2015) Difficulties and challenges related to the development of Homestay tourism in Sri Lanka

Chapter 10
Tourism Carrying Capacity and Destination Development

10.1 Introduction

The concept of Tourism Carrying Capacity (TCC) was developed in the 1980s (Elena and Franco 2010). It is related to the number of individuals, the available resources found in tourist places, and their assessment in the process of integrated planning and management, which is necessary for sustainable tourism development (UNEP/MAP/PAP 1997). Tourism activities largely depend on the carrying capacity of tourists' centers and their sustainable development. Although, tourism has positive impacts on economic development yet, it also has negative impacts on the local environment and native culture. A destination development approach based on the carrying capacity of the tourists/pilgrims' centers will manifest tourism development in the Himalayan region. The basic principle of sustainable tourism development lies in the preservation of ecological, socio-demographic, and economic-political dimensions, where the presence of human activities and processes represents the key factors (Hall 2011; UNWTO 2005). TCC is essentially an attempt to define the level of tolerance or compatibility between tourist activities and demands and the ecological, cultural, and economic support systems to meet the demands. In ecological terms, tourism activities have to be compatible with the maintenance and enhancement of the ecological balance; in social and cultural terms, it is compatible with the culture and the values of the people; and in economic terms, it needs to facilitate a process of development (WCED 1987). However, in the end, carrying capacity estimates often depend on administrative decisions about approximate sustainable levels of resource use (WTO/UNEP 1992). TCC analysis is necessary for the development of tourism, particularly in the mountainous regions where infrastructural facilities are comparatively less and tourists' inflow is high. Often, it leads to degradation of landscape and scarcity of accommodation and food. TCC represents a problem of allocation of scarce resources, e.g. protected natural or historical areas, to recreational opportunities that are density-dependent (McCool and Lime 2000). Destination development

V. P. Sati, *Sustainable Tourism Development in the Himalaya: Constraints and Prospects*, Environmental Science and Engineering, https://doi.org/10.1007/978-3-030-58854-0_10

implies a long-term process of change management from all levels of business in tourism to enhance the quality of life of local populations and the preservation of the cultural identity of tourist destinations. It also puts a way forward to the optimal economic development of destinations, increasing living standards, conservation of social and cultural heritage, and ecological conservation (Pearce 2015). Destination development of tourist places needs an organizational network for using common resources with common management of all segments.

It also performs activities at the micro-regional level, where all the stakeholders have individual and organizational responsibilities. Four basic interests and influential groups in tourism are government, the industry of entrepreneurs, tourists, and the local population (Byrd et al. 2009; Conaghan et al. 2010). The role of the government, which is the key holder of tourism development, is often emphasized within the concept of sustainable development based on the stakeholder approach (Hall 2011; UNWTO 2005). The Himalaya is world-famous for natural, cultural, aesthetic, and adventure tourism, which is being visited by several tourists and pilgrims from the ancient period. Practicing tourism in the Himalaya has been considered as a socioeconomic activity as tourism is the major source of income to the local people and revenue for the government. Further, a pilgrimage to the Himalaya is a centuries-old practice to the followers of Hinduism, which is deeply rooted in the culture and society (Sati 2015). With an increase in population within and outside of the region, there was a large increase in tourists and pilgrims' inflow in the Himalaya during the recent past. Although, a substantial increase in infrastructural facilities has been observed yet, it is not enough to meet the tourists' requirements. The Himalaya Mountain characterizes fragile and eco-sensitive landscapes. The tourists' centers and pilgrimages either are located along the river valleys or on the fragile slopes of the highlands, further accentuating fragility and vulnerability of landscapes. Carrying capacity of these destinations in terms of infrastructural facilities (mainly transportation and accommodation) to the tourists and pilgrims is insufficient and the development of these centers could not shape properly, and then environmental and ecological problems arose in the whole region. The development of most of the tourist and pilgrimage centers in the Himalaya is underneath. These centers lack proper and sufficient accommodation, toilets, transportation, drainage systems, parking, power supply, and food. Further, tourism is mostly practiced during the summer season when many tourists and pilgrims visit the whole region; the problems of infrastructural facilities are aggravated. In the meantime, several tourism/pilgrimage circuits have been developed. Figure 10.1 shows the TCC model for the Uttarakhand Himalaya. Three types of carrying capacities—Environmental Carrying Capacity (ECC), Cultural and Economic Carrying Capacity (CECC), and Institutional Carrying Capacity (ICC) are described.

This chapter discusses the TCC—ECC, CECC, and ICC, and destination development of five tourism places. It further describes the positive and negative tourism carrying capacity of all of them.

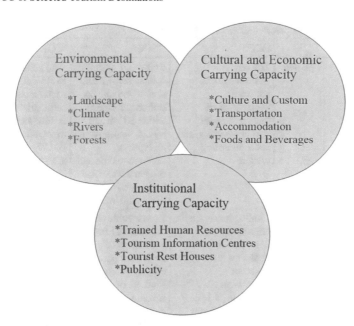

Fig. 10.1 Tourism carrying capacity

10.2 TCC of Selected Tourism Destinations

A case study of TCC of selected tourism destinations was conducted (Table 10.1). ECC, CECC, and ICC of Mussoorie, Gwaldam, Chakrata, Chopta, and Chilla were described.

10.3 Mussoorie

Mussoorie, the Queen of hills, and the honeymoon capital is situated at an altitude ranging between 1800 and 2000 m in the Middle Himalaya. It offers a scenic view of some of the Himalayan peaks—Bandarpunch, Chaukhamba, and Gangotri group of peaks. The famous Doon valley of the Uttarakhand can be viewed from Mussoorie. The holy river Ganga flows in the left and the Yamuna River flows in the west. It is one of the most popular hill stations of the Uttarakhand Himalaya and the country. A summer resort and a cultural center, Mussoorie has a spectacular landscape surrounded by lush green coniferous forests. Climate during the summer is quite feasible and during the winter, snowfalls, which attract tourists. It is 30 km from Dehradun, the state capital, and well connected by roads. Airport and railway station are located about 64 km and 30 km, respectively. It has several tourist locations that enhance TCC. Nearby Mussoorie, several natural and cultural tourist locales are

Table 10.1 TCC in the selected tourist/pilgrim places

Tourist places		ECC	CECC	ICC
Mussoorie	Positive	The serene environment, temperate vegetations, the feasible climate during summers, and snowfall during winters	Rich culture, well connected by road, airport and railway station is located within 30 km and 60 km, respectively, all types of accommodation and adequate water and sanitation facilities are available	Well popularized world-famous summer resort, substantial institutional supports, tour guides, and GMVN guest house are available
	Negative	Limited space, fragile terrain, high land degradation, forest depletion in the major trail routes, environmental pollution, garbage dumping problem	Roads within the town are narrow, congestion of roads during the peak tourism season, shortage of accommodation, cultural erosion	Tourism information centers are not much adequate
Gwaldam	Positive	The serene environment, pollution-free landscape, temperate vegetations, the feasible climate during summers, and snowfall during winters	The rich culture of Garhwal and Kumaon, abundance of traditional foods and beverages	GMVN guest house is available
	Negative	Limited space, expansion is not possible, sanitation, and garbage problem	Transportation facilities are poor; the road condition is very bad, inadequate accommodation	Institutional facilities are poor and inadequate, few tourists know about the locales
Chakrata	Positive	Spectacular landforms, temperate climate, summers are very feasible, snowfalls in winters, peaceful town, many tourist places lie in the surroundings, beautiful forest landscapes	Rich Jaunsari culture, enough traditional food and beverages	Well-known hill resort

(continued)

Table 10.1 (continued)

Tourist places		ECC	CECC	ICC
	Negative	Fragile landscape, water scarcity, sanitation problem, limited space	Poor transportation, bad road quality, inadequate accommodation	Institutional facilities are inadequate, lack in tourist information centers within the town, no trained guides
Chopta	Positive	It is alpine pastureland, the upper limit of temperate forest, spectacular landscape, feasible climate	Rich culture, enough traditional food and beverages	It has well known natural locales
	Negative	Fragile landscape, limited space	Poor infrastructural facilities—poor transport and accommodation	Institutional support is inadequate
Chilla	Positive	Well known eco-tourism destination, rich flora, and fauna	Close to Haridwar and Rishikesh cities	Forest department support for eco-tourism
	Negative	Fragile ecosystem	Inadequate accommodation facilities, poor transport facilities	Institutional support from the tourism department is poor

Source By author

situated, such as Dhanaulty and Surkanda. Trekkers start trekking to the Hari-Ki-Doon from Mussoorie. Mussoorie, established by Captain Young in 1825, was the summer capital of the British like Shimla in Himachal Pradesh. It slowly developed and is now considered an important tourist destination. It has modern bungalows, malls, and well-laid gardens. It has a range of quality hotels. Institutional facilities are quite adequate. It offers a beautiful nature walk known as Camel's Back Road. Gun Hill, Kempty Fall, Lal Tibba, and Mall Road are the major tourist destinations.

Despite having huge environmental, economic, and institutional tourism base, Mussoorie has many negative TCC. The landscape of Mussoorie is fragile. Landslides along the road routes during the monsoon season are frequent and intensive. Roads are narrow within the town, which leads to road congestion and consequently traffic jams during the peak tourism season.

10.4 Gwaldam

Gwaldam is a small and beautiful village, situated at 1950 m altitude, about 30 km from Kausani and about 52 km from Karnprayag. It is a border area of Garhwal and Kumaon region, located in the Chamoli district. A transitional zone of two different entities—the Garhwal Himalaya and the Kumaon Himalaya, Gwaldom has a mixed geo-environmental and cultural landscape, and therefore it obtains a rich culture and cultural heritage. It further comprises rich environmental components as lush green and dense coniferous forests, apple orchards, and tea gardens. The mighty Himalaya Mountain with Mount Nanda Devi, Trishul, and Nandaghunti lies close to Gwaldam enhancing its beauty and ECC. The Pindar valley's towns—Narayanbagar and Tharali and the Gomti valley's towns such as Baijnath and Bageshwar are located very close to Gwaldam, which provide a tourism support system. Gwaldam is the base for major trekking routes such as Bedni Bugyal and Roopkund and is known as a sleepy village.

The economic development of Gwaldam could not take shape, the first reason being its location. It is surrounded by dense vegetation and limited revenue land. Because of the rigorous implementation of the forest and wildlife acts of 1982, the sprawl of Gwaldam could not be possible. Accommodation facilities are inadequate. Few hotels and restaurants are available, which are not sufficient to accommodate tourists during the peak tourism season although, a tourist lodge of the tourism department and a forest bungalow are located here. Insufficient water supply and improper sanitation hamper tourism development. Because of the limited accommodation, few tourists stay in Gwaldam overnight. Transportation facilities are poor and road quality is bad. All these constraints have adversely impacted tourism development in Gwaldam.

10.5 Chakrata

Chakrata, a beautiful hill station, is known as the magical town. It is situated at 2200 m, on a ridge, about 98 km from Dehradun city. An administrative headquarters of Chakrata sub-division, it is located in Dehradun district. With dense coniferous forests, closeness to mighty Himalaya, and beautiful landscape, Chakrata town has rich environmental conditions. It has a variety of wildlife and birds, such as panthers, spotted dears, and wild fowls. Tiger fall is one of the most beautiful tourist destinations. Kharamba peak is the highest spot, a tourist destination, situated in the vicinity of the town. Summer is pleasant and snowfalls in winter provide a beautiful base for tourists to stay. It has rich Jaunsari tribal culture, quite different than the other parts of the Uttarakhand Himalaya. The Yamuna River flows close to it. The town was developed by the British East India Company. It has a cantonment area and because of it, the town and its surrounding areas are neat and clean.

Along with substantial TCC, Chakrata has several negative tourism aspects. It has a fragile landscape with limited cityscape. Because of the cantonment area and reserved forest in its surrounding areas, cityscape sprawl is not possible. Therefore, the construction of hotels, motels, and restaurants is not permitted, and they are limited. During the peak season of tourism, the tourists have to stay in other nearby small towns. Transportation facilities are limited. Only road transport is available. The quality of the road is bad. Roads are narrow and during monsoon season, landslides take place along the road, which hampers tourists' mobility. Institutional facilities are lagging.

10.6 Chopta

Chopta, a small village/settlement, is situated at an altitude of 2680 m in NDBR in Rudraprayag district, 29 km from Ukhimath. It is an alpine pastureland, the upper limit of coniferous forests. Rich faunal and floral diversity is found here. It is home to the famous musk dear. About 3 km upslope, a famous Hindu pilgrimage 'Tungnath' is situated. The pilgrims visiting Badrinath or Kedarnath stay overnight here. ECC is very rich in Chopta. Climate during the summer season is feasible whereas, during winter, snowfalls, therefore, it is called the 'Switzerland of Chamoli' district. The snow-clad mountain peaks are quite visible from Chopta. It is a major trekking route. A spectacular view of the mighty Himalaya can be seen from Chopta.

Despite rich ECC, Chopta has a very limited number of hotels. Since it is located within a wildlife sanctuary, it does not have electricity. Only solar power is available. A government guest house is located in Duggalbitta. The landscape is panoramic however, fragile. Thus, the construction of infrastructural facilities is not viable. Homestay can be provided in nearby villages located on the way between Ukhimath and Gopeshwar.

10.7 Chilla

Chilla is a wildlife sanctuary located within the RJNP. Lies in Pauri district, it has been developed as an eco-tourism destination. Chilla has rich floral and faunal diversity. It has more than 400 birds and wildlife species. Except for monsoon season, the entire year is suitable for eco-tourism. The Suswa River flows through the Chilla wildlife sanctuary, which is a tributary of the Ganga River. Chilla is located 8 km from Haridwar and 21 km from Rishikesh. A tourist rest house and a forest rest house are located here. A wildlife safari is practiced, which is about 50 km long trek and it takes about 2 h to complete a trek. Elephants, spotted deer, stag deer, bull, wild beer, fox, porcupine, jungle fowls, and peacocks are the main wildlife found in this sanctuary.

Chilla has rich ECC however, economic and ICC are lagging. Infrastructure—transportation and accommodation facilities are poor. Because this is a conserved and eco-sensitive area, the development of infrastructural facilities is not possible. In the meantime, Haridwar and Rishikesh cities are located close to Chilla therefore eco-tourism can be practiced. One day time, from morning to evening, is enough to visit the park. Haridwar and Rishikesh have a range of accommodation facilities, therefore, Chilla can be developed as the major eco-tourism destination.

10.8   Destination Development

Destination development is a catchy term and an important aspect of sustainable tourism. It includes proper transportation, adequate accommodation, ample water supply, waste and sanitation management, toilet facilities, communication facilities, and adequate food and beverages. Most of the tourism destinations in the Uttarakhand Himalaya have inadequate facilities. They are remotely located, the condition of transportation facilities are poor, and the quality of roads is bad. Water scarcity prevails mainly during the summers, which is the peak tourism season. Waste disposal and sanitation problems are acute. Except for Mussoorie, all four case study tourism destinations have poor infrastructural facilities. For sustainable destination development, all infrastructural facilities—transportation, accommodation, waste management, pure and ample water, and proper communication facilities are needed to be developed properly and adequately.

10.9   Conclusions

This study reveals that the TCC of tourist places has both positive and negative aspects. On the one hand, environmental carrying capacity is enormous as these tourist places have a spectacular landscape, feasible climate, rich biodiversity—faunal and floral, and rich culture and cultural heritage; on the other, economic and institutional carrying capacity is inadequate. Transportation and accommodation facilities are scarce, sanitation and waste management are poor, and landscape fragility is high. Most of the tourist places are remotely located where TCC is poor.

References

Byrd FJ, Bosley HE, Dronberger MG (2009) Comparison of stakeholder perceptions of tourism impact in rural eastern North Carolina. Tourism Manage 30(5):693–703

Conaghan, A., Hanrahan, J., Sligo, I.T.: Demand for and perceptions of sustainable tourism certification in Ireland. Retrieved from https://www.shannoncollege.com/wpcontent/uploads/2009/12/THRIC-2010-Full-Paper-A.-Conaghan-and-J.- Hanrahan.pdf (2010)

Elena M, Franco LF (2010) The carrying capacity of tourist destination: the case of a coastal Italian city, conference paper. https://www.researchgate.net/publication/230793635

Hall CM (2011) Policy learning and policy failure in sustainable tourism governance: From first-and second-order to thirdorder change? J Sustain Tourism 19(4–5):649–671

McCool SF, Lime DW (2000) Tourism carrying capacity: tempting fantasy or useful reality? J Sustain Tourism 9(5):372–388

Pearce DG (2015) Destination management in New Zealand: structures and functions. J Destination Market Manage 4(1):1–12

Sati VP (2015) Pilgrimage tourism in mountain regions: socioeconomic implications in the Garhwal Himalaya. South Asian J Tourism and Heritage 8(1):164–182

UNEP/MAP/PAP (1997) Guidelines for carrying capacity assessment for tourism in Mediterranean Coastal Areas, priority action programme, Regional Activity Centre, Split

UNWTO (2005) Making tourism more sustainable: a guide for policymakers. United Nations Environmental Programme, Paris

World Commission on Environment and Development (WCED) (1987) Our common future. Oxford University Press, Oxford

World Tourism Organization (WTO) and UNEP (1992) Guidelines: development of national parks and protected areas for tourism. WTO and UNEP, Madrid

Chapter 11
Sustainable Tourism Development: Constraints and Prospects

11.1 Introduction

Tourism is the major economic activity in the Uttarakhand Himalaya as it shares 50% of the total SGDP (UTDB 2019). Further, Uttarakhand has substantial potential for tourism development. However, tourism potential is not harnessed optimally. The economic development varies from the mountainous districts to the plain districts. About 20% of its population is living below the poverty line. Ecologically, the Uttarakhand Himalaya is very fragile. It is also vulnerable to natural disasters. Although tourism activities support economic development and enhance rural livelihoods yet, environmental implications as land degradation, forest depletion, and environmental pollution are enormous (UNEP 1999; ASEAN 1997). Sharma (1998) stated that tourism activity is important for economic development in the Himalayan region however, it has negative impacts on ecology and the environment. The conflict between ecology and economy is unresolved in the Himalaya (Brugger 1984). Here, environmental degradation in the highland pilgrimages is high due to large tourism activity (Bernbaum 1997). Tourism also has a significant impact on the mountain environment and downstream water supply (Bandyopadhyay et al. 1997; Buckley 1994). However, it is a very unevenly distributed phenomenon especially in the Himalayan river valleys (Sharma 1998).

The Uttarakhand Himalaya has both constraints and prospects for sustainable tourism development. On the one hand, it has enormous prospects for sustainable tourism development such as world-famous pilgrimages, rich culture and customs, world-class summer resorts, spectacular landscapes, varied and suitable climatic conditions, and welcoming nature of the native people, on the other, it has many constraints. For example, natural hazards and disasters, unplanned infrastructural development, mismanagement during the peak pilgrimages season, lack in parking facilities, bad quality of roads, potable water problem, sanitation problem, solid waste management problem, lack of clean and adequate public convenience facilities, lack

V. P. Sati, *Sustainable Tourism Development in the Himalaya: Constraints and Prospects*, Environmental Science and Engineering, https://doi.org/10.1007/978-3-030-58854-0_11

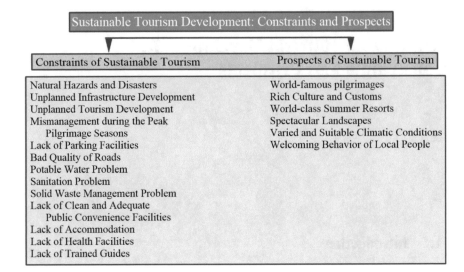

Fig. 11.1 Sustainable tourism development—constraints and prospects. *Source* By the author

of accommodation, lack of health facilities, and lack of trained guides are the major constraints (Fig. 11.1).

In this chapter, all these constraints and prospects of tourism development in the Uttarakhand Himalaya are discussed. The description of constraints and prospects are based on the author's knowledge about the region, data that are analyzed in the previous chapters, and bibliographic reviews.

11.2 Constraints of Sustainable Tourism

The landscape of the Uttarakhand Himalaya is vulnerable to climate-induced natural hazards and disasters (Sati 2014). It is an ecologically fragile, geologically sensitive, and tectonically and seismically active zone. The tourist places and pilgrimages are situated remotely and in the fragile landscapes. The natural locales are approached by road routes and by trekking. Road construction along the fragile slopes is not feasible. Huge landslides occur during the monsoon season, which is the main tourism season in the Uttarakhand Himalaya. The debris flows through the fragile slopes and deposits on the roadside, causing roadblocks. In the highland pilgrimages, large-scale land degradation occurs. Climate-induced natural disasters are very common. Cloudburst triggered debris flows and flash floods are common during the monsoon season (Sati 2018). Since the region receives heavy downpour, the streams and rivers inundate and flow over danger marks. Every year, several incidences of debris flows and flash floods occur, which results in the death of tourists and pilgrims. In June 2013, the entire Uttarakhand Himalaya received heavy rain. All the major rivers were inundated. This gave birth to the Kedarnath and Badrinath Tragedy, which is called the 'Himalayan Tsunami'. This incident resulted in the deaths of more than 10,000 tourists/pilgrims and thousands went missing. The settlements and agricultural land along the rivers were washed away, which led to huge economic loss (Sati 2013).

The infrastructural facilities are lagging and whatever the facilities are available, they are not maintained and used properly. Similarly, tourism development is also unplanned. Tourism is practised mainly from May to October, which is called the peak tourism season. A huge number of tourists/pilgrims visit the tourist places and pilgrimages during the season however, there's no propoer management. The tourist places are overcrowded during the time. The carrying capacity of these tourist places and pilgrimages is not adequate, therefore the tourists/pilgrims face enormous problems such as accommodation and transportation. Further, parking facilities in the tourist places are inadequate therefore, the tourists/pilgrims come back from the highland pilgrimages/tourist places on the same day. Many tourist places are remotely located where the tourist/pilgrim visit by trekking. The quality of the roads is very poor. Many mountainous roads are narrowly constructed along the fragile slopes and the river valleys. During the rainy season, the quality of roads gets further deteriorated. Potable water scarcity prevails in many tourist places/pilgrimages although water availability is ample. Figure 11.2 shows a huge landslide near Sakinidhar on the way to Badrinath and Kedarnath (a) and enormous debris above the Kedarnath pilgrimage, causing severe land degradation (b).

Sanitation and dumping of solid waste are the biggest problems. Many tourist places/pilgrimages are constructed without any planning. They are old cities/towns, where proper sanitation work is not carried out. Further, increasing population and mass tourism in these cities and towns lead to dumping of more solid waste. Because there is no proper waste management system in these places, environmental pollution has become a major problem. Public convenience facilities are not adequate, and existing one are not cleaned. During the monsoon season, when this region receives heavy rainfall, the street water is over-flooded because of poor drainage.

Fig. 11.2 **a** a huge landslide between Rishikesh and Devprayag, near Sakinidhar, **b** huge debris above the Kedarnath pilgrimage

(a)   (b)

Fig. 11.3 a Traffic congestion near Joshimath town because of the narrow road (b) water pollution in the Mandakini River, near Gaurikund (close to source area)

The flowing of waste materials on these streams leads to severe water pollution situation. Bus stops, public pedestrians, and green belts are not adequate. Whenever and wherever, they are available, they are not up to mark position therefore they lead constraints to tourists/pilgrims. Accommodation is not adequate. Further, the variety of accommodation is not sufficient. As the pilgrims visiting Uttarakhand mainly belong to medium and lower-income levels, they do not find reasonable and comfortable accommodation. Health facilities are lagging. Many tourist places are remotely located and some are inaccessible, where health facilities are not substantial. Further, the climatic conditions are very harsh in the highland pilgrimages and tourist places. During the tourism season, heavy rainfalls occur, which leads to health-related problems. Few hospitals are available however facilities in these hospitals are not adequate. Although, education level in the Uttarakhand Himalaya is high yet the educated youth out migrate to other parts of the country. Therefore, trained tourist guides are largely unavailable. The foreigners face language problems in the tourists/pilgrims places. Figure 11.3 shows traffic congestion near Joshimath town because of the narrow road (a) and severe water pollution by the dumping of solid waste in the Mandakini River near Gaurikund (b). The place is very close to the source area of the river and there are no settlements along the river. Dumping of waste mainly non-biodegradable material, is due to mass tourist/pilgrim mobility.

11.3 Prospects of Sustainable Tourism Development

Besides many constraints of tourism development in the Uttarakhand Himalaya, there are several prospects for sustainable tourism development. The study area has three-dimensional landscapes, from the valley regions to the Middle Himalaya and the Great Himalayan Ranges, which make a panoramic river topographies, mountain landscape, and glacial topographies. These features provide a suitable base for tourism development. Among the major drivers and prospects of sustainable tourism in the Uttarakhand Himalaya, world-famous pilgrimages, rich culture and customs, world-class summer resorts, spectacular landscapes, varied and suitable climate, and welcoming behavior of local people are prominent.

One of the major prospects of tourism development in the Uttarakhand Himalaya is the highlands and the river valleys pilgrimages. The Uttarakhand Himalaya is known for its pilgrimages, where exodus number of pilgrims visit every year. Pilgrimages to the Himalaya is a century-old practice. The present study shows that although the highland pilgrimages have equal importance as the valley pilgrimages yet, the pilgrims visiting the valley pilgrimages are outnumbered because the valley pilgrimages are well connected by all means of transportation—roads, airways, and railways. Further, other infrastructural facilities are also substantial in the river valley pilgrimages. This part of the Himalaya has a rich culture and customs. Fairs and festivals are celebrated every month. There are several festivals of India, which are celebrated only in the Uttarakhand Himalaya. They are localized. Among them, the major fairs and festivals are Baishakhi, NDRJ, Kumbh (haft and full), Ganga Dussehra, Pandav Nritya, Sakrantis, Magh Mela, and Uttarayani. The food, beverages, clothing, folklore—songs and dances are different than the other parts of the Himalaya and country. The pilgrims—domestic and foreigners participate in these fairs and festivals and assist in tourism development.

In the Uttarakhand Himalaya, world-class summer resorts are situated. Some of them are Mussoorie, Nainital, Ranikhet, and Almora. The climate during the summer in these resorts is feasible whereas snowfalls during winters. In both seasons, tourist visits these resorts for leisure tourism. Tourists' inflow during summers is high because of the feasible climate. Infrastructural facilities in these resorts are substantial. A range of accommodation is available, including star hotels. The landscape is spectacular. The river valleys, Shivalik hills, Middle Himalaya, and the Great Himalayan ranges have unique topographies. Climatic conditions vary from the river valleys to the Middle Himalaya, and the Greater Himalaya. Tourism is favorable in the river valleys during winters. In the Middle Himalaya, it is suitable during the summer. Therefore tourism is practised both in summer and winter seasons. The local people are well behaved and innocent. They believe the Guest to be God (Athiti Devo Bhavah). All these prospects of tourism manifest sustainable tourism development in the Uttarakhand Himalaya.

11.4 SWOC Analysis

SWOC analysis which includes strengths, weaknesses, opportunities, and challenges was carried out in terms of tourism development in the Uttarakhand Himalaya (Fig. 11.4). The major strengths of tourism development in the Uttarakhand Himalaya are unique mountain environment of the Himalaya, spectacular landscapes—geographical features, presence of the Himalayan mountain villages and

Strengths	Weaknesses
• Unique mountain environment of the Himalaya • Spectacular landscapes – geographical features • Presence of the Himalayan mountain villages and towns • Mountain peaks • Trekking paths and circuits • National parks and wildlife sanctuaries • Conservation of natural areas	• Unplanned development • Prone to natural disasters • Poor urban environment – high pollution • Lacking publicity of the natural environment • Lacking awareness about the potential of ecotourism • Lacking tourism facilities • Lacking coordination among the government agencies • Poor transportation • Poor traffic management • Poor participation of the community in tourism development • Local people not being benefited from tourism output

SWOC ANALYSIS

Opportunities	Challenges
• High potential for ecotourism and responsible tourism • High biodiversity • Abundant adventure tourism activities • High potential of water sports • High potential of rural and cultural tourism • Employment opportunities • Healthy environment • Potential of holiday homes • Adequate education facilities	• The occurrence of natural disasters • Uneven and irregular tourists/pilgrims' inflow • Unorganized tours and travels • High traffic and pollution • Poor maintenance of roads

Fig. 11.4 SWOC analysis of sustainable tourism development in the Uttarakhand Himalaya

towns, mountain peaks, trekking paths and circuits, national parks and wildlife sanctuaries, and conservation of natural areas. Similarly, there are many weaknesses, which hamper tourism development such as unplanned development, the landscape being highly prone to natural disasters, poor urban environment—high pollution, lacking publicity of the natural environment, lacking awareness about the potential of ecotourism, lacking tourism facilities, lacking coordination among the government agencies, poor transportation, poor traffic management, poor participation of the community in tourism development, and local people not being benefited from the tourism output.

The third component of SWOC analysis is opportunities, which include high potential of ecotourism and responsible tourism, high biodiversity, abundant adventure tourism activities, high potential of water sports, high potential of rural and cultural tourism, employment opportunities, health environment, potential of holiday homes, and adequate education facilities. Challenges are the fourth and the last component of SWOC analysis. The occurrence of natural disasters, uneven and irregular tourists/pilgrims' inflow, unorganized tours and travels, high traffic and pollution, and poor maintenance of roads are the major challenges.

11.5 Conclusion

This study reveals that the Uttarakhand Himalaya has both constraints and prospects of tourism development. On the one hand, it has lots of constraints therefore the tourism development could not take shape substantially; on the other, it has plenty of prospects for tourism development. These prospects can be harnessed optimally through framing appropriate policy measures and implementing them. Most of the constraints are manmade; they can be turned into prospects. The government agencies should work together with the community people. Maintenance of roads, waste management, development of support system, along with providing basic facilities to the tourists/pilgrims will manifest sustainable tourism development in the Uttarakhand Himalaya.

References

ASEAN (1997) Economic co-operation: transition and transformation. Institute of Southeast Asian Studies

Bandyopadhyay J, Rodda JC, Kattlemann R, Kundzewicz ZW, Kraemer D (1997) Highland waters—a resource of global significance. In; Messerii B, Ives JD (eds) Mountains of the world: a global priority. Parthenon Press, London, pp 131–156

Bernbaum E (1997) The spiritual and cultural significance of mountains. In: Messerii B, Ives JD (eds) Mountains of the world: a global priority. Parthenon Press, New York, pp 39–60

Brugger EA (1984) In The transformation of Swiss Mountain Regions. In: Brugger EA, Furrer G, Messerii B, Messerii P (eds) Verlang Paul Haupt Bern, Switzerland

Buckley R (1994) A framework for ecotourism. Ann Tourism Res 21:661–667

Sati VP (2013) Extreme weather related disasters: a case study of two flashfloods hit areas of Badrinath and Kedarnath Valleys, Uttarakhand Himalaya, India. J Earth Sci Eng 3:562–568. ISSN: 2159–581X

Sati VP (2014) Landscape vulnerability and rehabilitation issues: a study of hydropower projects in the Garhwal region, Himalaya. Nat Hazards 75(3):2265–2278. https://doi.org/10.1007/s11069-014-1430-y

Sati, V.P. (2018) Cloudburst triggered natural hazards in Uttarakhand Himalaya: mechanism, prevention and mitigation. Int J Geol Environ Eng 12(1):38–45

Sharma P (1998) Sustainable tourism in the Hindukush-Himalaya: issues and approaches. In: East P, Luger K, Inmann K (eds) Sustainability in mountain tourism: perspectives for the Himalayan countries. Book Faith India, Delhi and Studienverlag, Innsbruck, pp 47–69

UNEP (1999) The global importance of tourism. World Travel & Tourism Council and International Hotel & Restaurant Association, Background Paper #1, Commission on Sustainable Development, Seventh Session, 19–30 April 1999, U N Department of Economic and Social Affairs, New York

UTDB (2019) Annual report. Uttarakhand Tourism Development Board, Dehradun

Chapter 12
Conclusions

The description and analysis of tourism development in the Uttarakhand Himalaya show that it has enormous potential for sustainable tourism development, as rich geo-environment and cultural components provide a suitable base. It has been noticed that the Uttarakhand Himalaya has rich geographical attributes in the forms of natural landscapes, forest landscapes, river topographies, climate, and spectacular rural locales. These geographical attributes provide a suitable platform for tourism development. However, the natural beauty of many locales has not been harnessed optimally, because of several drawbacks such as lacking infrastructural facilities and institutional supports. The need of the hours is to develop the destinations for sustainable tourism development.

A study on cultural components of tourism reveals that the Uttarakhand Himalaya has a rich and unique culture and cultural heritage for sustainable tourism development. The pilgrimages are the major destinations for cultural, spiritual, and religious tourism. It has been noticed that pilgrims, who visit Uttarakhand outnumber the tourists. Rich culture, customs, fairs, and festivals of the Uttarakhand Himalaya are the major tourist attractions. The major types of tourism and tourist/pilgrim centers are described. A large proportion of visitors, more than 50% both domestic and foreign, visit for cultural activities including pilgrimages.

The nature of tourism and tourists/pilgrims' inflow has been described. Study on tourists/pilgrims' inflow depicted that pilgrims' inflow in the pilgrimages was higher than tourists' inflow in the natural locales. Further, the river valley pilgrimages were visited more by the pilgrims than the highland pilgrimages. The trend of tourists/pilgrims' inflow was not uniformed in all the pilgrimages and natural locales. The inflow decreased during the occurrences of natural disasters, mainly in the highland pilgrimages and natural locales. However, tourists/pilgrims' inflow increased continuously.

© The Editor(s) (if applicable) and The Author(s), under exclusive license
to Springer Nature Switzerland AG 2020
V. P. Sati, *Sustainable Tourism Development in the Himalaya:
Constraints and Prospects*, Environmental Science and Engineering,
https://doi.org/10.1007/978-3-030-58854-0_12

The Uttarakhand Himalaya has been divided into the major tourist/pilgrim circuits. These circuits are characterized as natural, cultural, park and wildlife, and adventure. Within the major tourist/pilgrim circuits, some small tourist circuits are situated. Out of them, few circuits are developed with substantial tourism infrastructural facilities. Many of them are remotely located with limited tourism facilities. Some pilgrim circuits are accessible through trekking only. In the meantime, the potential for sustainable tourism development in these circuits is enormous. These tourists/pilgrims' circuits can be developed by providing adequate tourism facilities.

Three tourism routes and their geographical and cultural importance were studied. These routes have a high potential for sustainable tourism development. In the meantime, some of the tourist locales on these routes are underdeveloped. Infrastructural facilities are lagging. Accessibility of some tourist places is very difficult, mainly in the highland pilgrimages. Accommodation and institutional facilities are not up to mark. Therefore, it has been noticed that the tourists/pilgrims' inflow in these places is considerably less whereas, the geographical and cultural bases of tourism locales are significantly high.

Infrastructural facilities—transportation, accommodation, and institutions, have been described. The roads are traversed in the river valleys and the Middle Himalaya whereas the highlands (Upper Himalayan region) are still inaccessible. Many tourist places are not connected by any means of transportation. Further, the condition of roads is critical, which leads to roadblocks and road accidents. The railway is only available in the plain regions. Further, the construction of the rail route along the fragile slope is not suitable. Air transportation is suitable; however, there are only five airports and a helipad, from which only Dehradun airport is connected with the major cities of India. Construction of roads is possible through pillars and tunnels, which will minimize roadblocks and road accidents. The major tourist places can be connected to the airways and the fare can be subsidized by the government to promote air transportation.

Accommodation facilities are not adequate. However, the tourists/pilgrims' inflow is very high. As a result, the tourists/pilgrims come back from the tourist places/pilgrimages on the same day. Homestay facilities can be developed in the rural and urban areas, where major tourist places and pilgrimages are located. The income level of tourists/pilgrims varies. Many of them have low-income levels therefore, a range of accommodation can be provided while keeping the income level of the pilgrims and tourists in mind. Institutional facilities need to be enhanced. There are several places of tourists' interests, which have not yet been explored. UTDB can play a vital role in developing lesser-known places with the help of GMVN and KMVN. If all the infrastructural facilities are developed and provided to tourists/pilgrims, then the Uttarakhand Himalaya may be a hub of tourism activities. Subsequently, it will lead to economic development through income generation and employment augmentation.

Homestay tourism is an emerging sector for economic growth and rural livelihood enhancement in the Uttarakhand Himalaya as it has significant prospects. The tourist places can be developed and the rich culture and delightful delicious cuisine of the state can be popularized through homestay tourism. The homestay tourism can also augment employment, make the rural people self-reliant, and can check/minimize the forced rural–urban migration. The homestay policy can act as a tool for developing Uttarakhand as a tourism state. Homestay may not be successful without government support, because it needs lots of infrastructural facilities. Therefore, the government should play a more rigorous role to develop homestay tourism. In Uttarakhand, the hotel industries are developed in major cities, tourist places, and pilgrimages, and owned by businessmen from within and outside the state. Rural areas and the poor local people, who are tourism service providers, are debarred from the benefits of tourism development. The development of homestay tourism will manifest strengthening rural economy and livelihood. The state government should ensure that the benefits of homestay tourism go to the local people.

Tourism carrying capacity has both positive and negative aspects. On one hand, environmental carrying capacity is enormous, as the tourist places have a spectacular landscape, feasible climate, rich biodiversity—faunal and floral, and rich culture and cultural heritage; on the other hand, economic and institutional carrying capacity are inadequate. Transportation and accommodation facilities are scarce, sanitation and waste management are poor, and landscape fragility is high. Most of the tourist places are remotely located where tourism carrying capacity is poor.

Constraints and prospects of tourism development in the Uttarakhand Himalaya have been described. On one hand, it has lots of constraints therefore, tourism development could not take place substantially; on the other hand, it has plenty of prospects for tourism development. The prospects can be harnessed optimally if appropriate development strategies are adapted. Most of the constraints are manmade; they can be turned into prospects. Proper policy framing and their implementation will manifest sustainable tourism development in the Uttarakhand Himalaya.

12.1 Recommendations

The study on sustainable tourism development in the Uttarakhand Himalaya reveals that the state has well-established pilgrimage sites and natural locales. However, there is still a high potential to develop destinations for spiritual, yoga, and health tourism. Further, the development of adventure tourism, such as trekking, mountaineering, skiing, and river rafting, along with rural and ecotourism have enormous scope in the Uttarakhand Himalaya. Sustainable tourism development can be achieved by providing basic amenities—transportation, accommodation, institutions, and health; skilled labours and service providers.

Theme based tourist circuits—natural, cultural, adventure, health, and park and wildlife—can be developed. The state also has spiritual and religious-based tourist circuits. All these circuits can be developed for sustainable tourism. The rural areas of the Uttarakhand Himalaya preserve the rich traditional culture, customs, and rituals, which attract tourists and pilgrims from the Indian subcontinent and abroad. Homestay in rural areas can supplement accommodation. The tourists and pilgrims can be acquainted with folk culture. Ecotourism has substantial prospects. All these bases can be linked to the development of tourism circuits. The development agencies of the state government can promote cleanliness and beautification of the rural areas for attracting tourists/pilgrims.

There is a need of classifying infrastructural gaps and solving them so that the potential of tourist circuits can be harnessed. An integrated and holistic approach can be developed for the sustainable development of tourist circuits. The state government has proposed to develop 15 new and lesser known destinations in 13 districts based on the tourism themes—natural, cultural, spiritual, adventure, park and wildlife, and ecotourism. It has also planned to develop the entire state so that it gets included among the top 10 tourism destination states of the country by 2020, and the top 3 destination states by 2023 (MoT 2019).

These 15 destinations are Munsiyari in Pithoragarh district, Kausani in Bageshwar district, Katarmal in Almora district, Mukteshwar in Nainital district, Lohaghat in Champawat district, Parag Farm in USN district, Chopta in Rudraprayag district, Tehri Lake in Tehri district, Chinyalisaur in Uttarkashi district, Gairsain and Auli in Chamoli district, Khirsu in Pauri district, Chakrata in Dehradun district, and Piran Kaliyar and Shakipeeth in Haridwar district. Some of them are classified as theme-based tourist/pilgrim destinations. For example, Munsiyari and Mukteshwar have been developed as lesser tourism, tea tourism is developed in Kausani, meditation in Katarmal, hill station in Lohaghat, and Parag Farm got amusement park. Chopta is developed as ecotourism, Tehri Lake as water sports, and Khirsu has been developed as a hill station and wildlife tourism.

Sustainable tourism development depends on the collaboration of all the stakeholders of tourism-related activities such as the state government, tourism departments at district and state levels, and local community people. Tourism infrastructure and the safety of tourists/pilgrims are the major drivers for sustainable tourism development. The state of Uttarakhand is fortunate enough to have a safe environment however, infrastructure has yet to be developed in the remote and inaccessible areas mainly along the trekking and mountaineering routes. In the meantime, the plain areas of the state are well connected with the other parts of the country and abroad by air, rail, and roadways. Besides the infrastructural facilities, there is a need to develop water and sanitation facilities, electricity, urban development, hospitality, IT, and disaster management facilities that can pave a way for sustainable tourism development.

A tourism development plan with coordination of the Government of India, the Government of Uttarakhand, the United Nations Development Programme (UNDP), and the World Tourism Organization has been developed for 2007–22, which is known as 'Uttarakhand Tourism Development Plan'. The main purpose of this plan is to develop high-quality sustainable tourism infrastructure, sustainable development of tourism destinations, and prime tourism zones. It also aims to identify the major tourism zones and develop them as tourists/pilgrims circuits. Besides, natural tourism resources can be developed. The Uttarakhand Himalaya has enormous tourism resources, which are largely unused. It attracts the tourists and pilgrims for natural tourism, cultural tourism, pilgrimages, adventure tourism, park and wildlife tourism, ecotourism, rural and health tourism. These bases of tourism support sustainable tourism development. The diverse tourism products such as natural beauty, unique Himalayan environment, as well as historical and cultural assets further enhance the potential of sustainable tourism in the Uttarakhand Himalaya.

Environmental and cultural assets/products are needed to be conserved for the sustainable development of tourism. Lacking infrastructure is one of the drivers that have constraints in tourism development. All the tourism circuits can be connected with a sustainable mode of transportation. Tourism in the Uttarakhand Himalaya has seasonality, meaning tourism is practiced in two seasons. The peak season is the summer season when the highland pilgrimages are opened for the pilgrims, drawn from different corners of the country. The other season is winter when snow falls heavily in the Middle Himalaya and the tourists visit for enjoying the snow-fall. The seasonality aspect should be taken into consideration when the policy for tourism development is framed and implemented. Along with practicing different types of tourism, culture, handicrafts, and local cuisine can be developed and preserved. Publicity and marketing of the tourists/pilgrims' centers are required to harness the huge tourism potential of the region. Local people can be trained for hosting the tourists/pilgrims. Further, trained tourist guides are needed for satisfying tourists/pilgrims while they are performing tourism in the region. The public and private sectors can work together to boost tourism development. The abundant tourism resources should be mobilized for sustainable tourism through the development of infrastructural facilities. The UTDB with the cooperation of GMVN and KMVN can develop an inclusive master plan that can integrate tourism resources with sustainable tourism development more comprehensively and holistically.

Both domestic and foreign tourists and pilgrims are increasing gradually as the trend of tourists/pilgrims' inflow shows. Further, it was also noticed that out of the total tourists/pilgrims' inflow, more than 50% of tourists/pilgrims visit the Haridwar pilgrimage only. The reason is that the mainland of the Uttarakhand Himalaya is lagging in suitable infrastructural facilities therefore the tourists/pilgrims visit mainly the river valleys tourist places and pilgrimages. The issue of infrastructural facilities should also be taken into consideration while framing and implementing sustainable tourism policy.

Awareness program on the development of tourism destinations through effective branding of the state, marketing strategy, and private–public partnership will enhance tourism development. There are some globally important tourist destinations such as Haridwar and Rishikesh—important for religious tourism, Dehradun the capital city, Nainital and Mussoorie the hill resorts, USN known for a commercial visit, and CNP famous for tiger reserve. These centers have substantial tourism infrastructural facilities and can be further developed for sustainable tourism development. The international tourism market for the tourists/pilgrims has to be provided. This will also manifest the economic development of the state. Development of transportation, electricity, drinking water supply, telecommunications, emergency services, restaurants and hotels, and waste disposal should be ensured. Transport linkages, within and outside the state can be developed. Cable cars and ropeways can be constructed in some of the important tourist destinations, where transportation facilities are not much available. Tehri Lake can be developed for water sports and car and passenger ferry services can be introduced.

Cleanly environment can be provided to tourists/pilgrims. In tourist places, clean toilets should be provided after every km, which is lacking in many tourist centres in Uttarakhand. Cleanliness drive is also a component of medical/health tourism. The Uttarakhand Himalaya has the potential for developing medical/health tourism therefore cleanliness drive is an important aspect of sustainable tourism development. Land degradation is the biggest problem in the tourist places of Uttarakhand. Soil erosion along the trails in the highland pilgrimages and natural places is an enormous problem. Trail quality can be improved so that soil erosion can be checked. Many nature lovers are interested in a nature walk therefore, walkways can be created and constructed.

NITI Aayog has addressed 12 core areas of current policies and plans for sustainable tourism practices (NITI Aayog 2018). These areas are disaster management, pollution control, visitor control, tourism traffic management, crisis management, waste management, natural resources, and environmental management, quality standard/control mechanism, tourism enterprise development governance, energy, gender, and marketing and branding. Policy measures to address these core areas are required for tourism development in the state.

Although, hotels and resorts are mushrooming yet, the local people are not benefitted because these hotels and resorts are owned by the people who do not belong to the region. The local people are largely engaged in practicing agriculture. As the output from agricultural products is very low, the youth of the region out-migrate in the search for jobs. The development of infrastructural facilities along the major roads and in the tourist places will further support tourism development. Practicing ecotourism and providing a homestay facility will enhance the income and economy of the local people. Safety and security measures in tourist places and along the roads need to be framed and implemented rigorously so that tourists can enjoy tourism without any fear. Tourism needs adequate infrastructural facilities such as parking facilities, trained and skilled tourism personnel, and medical facilities besides others.

Four stakeholders can assist in developing tourism, which are (a) the tourists, (b) the developers (c) providers of services, and (d) the local perspective. Ecotourism is a more suitable environmental point of view than mass tourism. Ecotourism has two modern trends—integration and conservation with development and qualitative change in tourism demand.

Natural and cultural tourism in the Uttarakhand Himalaya is a centuries-old practice. However, ecotourism, health/medical tourism, and rural/village tourism are yet to be initiated, which have significant potential in the Uttarakhand Himalaya. If all the suggestions are accepted and policies on them are framed and implemented, the state of Uttarakhand can attain sustainable tourism development. It will lead to income generation, employment augmentation, livelihood enhancement, and economic development of the people and the state. And finally, it will check/minimize out-migration, contributing to making the state prosperous.

References

MoT (2019) Ministry of Tourism and Government of India, New Delhi
NITI Aayog (2018) Report of Working group II sustainable tourism in the Indian Himalayan region, New Delhi

Index

Printed in the United States
by Baker & Taylor Publisher Services